PROOFS AND REFUTATIONS

FOR RETURN

PROOFS AND REFUTATIONS
The Logic of Mathematical Discovery
IMRE LAKATOS

Edited by
John Worrall and Elie Zahar

CAMBRIDGE
UNIVERSITY PRESS

CAMBRIDGE UNIVERSITY PRESS
Cambridge, New York, Melbourne, Madrid, Cape Town, Singapore, São Paulo

Cambridge University Press
The Edinburgh Building, Cambridge CB2 2RU, UK

Published in the United States of America by Cambridge University Press, New York

www.cambridge.org
Information on this title: www.cambridge.org/9780521210782

First published 1976
Reprinted with corrections 1977, 1979, 1981
Reprinted 1983, 1984, 1987, 1988, 1989, 1991, 1993, 1994, 1995,
1997, 1999

A catalogue record for this publication is available from the British Library

ISBN-13 978-0-521-29038-8 paperback
ISBN-10 0-521-29038-4 paperback

Transferred to digital printing 2005

CONTENTS

CONTENTS

EDITORS' PREFACE

Our great friend and teacher Imre Lakatos died unexpectedly on 2 February 1974. At the time he was (as usual) engaged on many intellectual projects. One of the most important of these was the publication of a modified and extended version of his brilliant essay 'Proofs and Refutations', which appeared in four parts in *The British Journal for the Philosophy of Science*, **14**, 1963–4. Lakatos had long had a contract for this book, but had held back publication in the hope of amending and further improving the essay, and of adding to it substantial extra material. This work was considerably delayed by the diversion of his interests to the philosophy of physical science, but in the summer of 1973 he finally decided to go ahead with the publication. During that summer we each discussed plans for the book with him, and we have tried to produce a book which, in the sadly changed circumstances, is as similar as possible to the one then projected by Lakatos.

We have thus included three new items in addition to the original 'Proofs and Refutations' essay (which appears here as Chapter 1). First we have added a second part to the main text. This concerns Poincaré's vector-algebraic proof of the Descartes–Euler conjecture. It is based on chapter 2 of Lakatos's 1961 Cambridge Ph.D. thesis. (The original 'Proofs and Refutations' essay was a much amended and improved version of chapter 1 of that thesis.) A part of chapter 3 of this thesis becomes here appendix 1, which contains a further case-study in the method of proofs and refutations. It is concerned with Cauchy's proof of the theorem that the limit of any convergent series of continuous functions is itself continuous. Chapter 2 of the main text and appendix 1 should allay the doubt, often expressed by mathematicians who have read 'Proofs and Refutations', that, while the method of proof-analysis described by Lakatos may be applicable to the study of polyhedra, a subject which is 'near empirical' and where the counter-examples are easily visualisable, it may be inapplicable to 'real' mathematics. The third additional item (appendix 2) is also based on a part of chapter 3 of Lakatos's thesis. It is about the consequences of his position for the development, presentation and teaching of mathematics.

One of the reasons Lakatos delayed publication was his recognition that some of this extra material, whilst containing many new points and developments of his position, was in need of further consideration and further historical research. This is particularly true of the material (in appendix 1) on Cauchy and Fourier. We also are aware of certain difficulties and ambiguities in this material and of omissions from it. We felt, however, that we should not change the content of what Lakatos had written. As for elaborating on, and adding to, the material, neither of us was in a position to supply the necessary long and detailed historical research. Faced then with the alternatives of not publishing the material at all, or publishing it in an unfinished state, we decided on the latter option. We feel that there is much of interest in it, and hope that it will stimulate other scholars to extend and correct it if necessary.

In general, we did not think it right to modify the content of Lakatos's material, even those parts of it about which we were confident Lakatos had changed his position. We have therefore restricted ourselves to pointing out (in notes marked with asterisks) some of those things we should have tried to persuade Lakatos to change and (which often amounts to the same thing) some of those points we believe Lakatos would have changed in publishing this material now. (His intellectual position had, of course, changed considerably during the thirteen years between completing the Ph.D. thesis and his death. The major changes in his general philosophy are explained in his [1970]. We should mention that Lakatos thought that his methodology of scientific research programmes had important implications for his philosophy of mathematics.)

Our approach to matters of presentation has been to leave the material which Lakatos had himself published (i.e. chapter 1 of the main text) almost entirely unchanged (the only exceptions are a few misprints and unambiguous minor slips). We have, however, rather substantially modified the previously unpublished material – though, to repeat, only in form and not in content. Since this may seem a rather unusual procedure, perhaps a few words of justification are in order.

Lakatos always took a great deal of care over the presentation of any of his material which was to be published, and, prior to publication, he always had such material widely circulated amongst colleagues and friends, for criticism and suggested improvements. We are sure that the material here published for the first time would have undergone this treatment, and that the changes would have been more drastic than those we have dared to introduce. Our knowledge (through personal experience) of the pains Lakatos took to present his position as clearly as possible obliged us to try to improve the presentation of this material as best we could. It is certain that these new items do not read as well as they would have done, had Lakatos himself revised the material on which they are based, but we felt that we were close enough to Lakatos, and involved enough in some of his previous publications, to make a reasonable attempt at bringing the material up to somewhere near his own high standards.

We are very pleased to have had the opportunity to produce this edition of some of Lakatos's important work in the philosophy of mathematics, for it allows us to discharge part of the intellectual and personal debt we both owe him.

John Worrall
Elie Zahar

ACKNOWLEDGMENTS

The material on which this book is based has had a long and varied history, as is in part already indicated in our preface. According to the acknowledgments Lakatos appended to his original essay of 1963–4 (reprinted here as chapter 1), that work began life in 1958–9 at King's College, Cambridge, and was first read at Karl Popper's seminar at the London School of Economics in March 1959. Another version was incorporated in his 1961 Cambridge Ph.D. thesis, on which the rest of this book is also based. The thesis was prepared under the supervision of Professor R. B. Braithwaite. In connection with it, Lakatos acknowledged the financial assistance of the Rockefeller Foundation and that he 'received much help, encouragement and valuable criticism from Dr T. J. Smiley'. The rest of Lakatos's acknowledgments read:

When preparing this latest version at the London School of Economics the author tried to take note especially of the criticisms and suggestions of Dr J. Agassi, Dr I. Hacking, Professors W. C. Kneale and R. Montague, A. Musgrave, Professor M. Polanyi and J. W. N. Watkins. The treatment of the exception-barring method was improved under the stimulus of the critical remarks of Professors G. Pólya and B. L. Van der Waerden. The distinction between the methods of monster-barring and monster-adjustment was suggested by B. MacLennan.

The paper should be seen against the background of Pólya's revival of mathematical heuristic, and of Popper's critical philosophy.

The original essay of 1963–4 carried the following dedication:

For George Pólya's 75th and Karl Popper's 60th Birthday.

In preparing this book, the editors were helped by John Bell, Mike Hallett, Moshé Machover and Jerry Ravetz, who all kindly read drafts of chapter 2 and the appendices, and produced helpful criticisms.

We should also like to acknowledge the work of Sandra D. Mitchell and especially of Gregory Currie, who carefully criticised our reworking of Lakatos's material.

J. W.
E. Z.

AUTHOR'S INTRODUCTION

It frequently happens in the history of thought that when a powerful new method emerges the study of those problems which can be dealt with by the new method advances rapidly and attracts the limelight, while the rest tends to be ignored or even forgotten, its study despised.

This situation seems to have arisen in our century in the Philosophy of Mathematics as a result of the dynamic development of metamathematics.

The subject matter of metamathematics is an abstraction of mathematics in which mathematical theories are replaced by formal systems, proofs by certain sequences of well-formed formulae, definitions by 'abbreviatory devices' which are 'theoretically dispensable' but 'typographically convenient'.[1] This abstraction was devised by Hilbert to provide a powerful technique for approaching some of the problems of the methodology of mathematics. At the same time there are problems which fall outside the range of metamathematical abstractions. Among these are all problems relating to informal (*inhaltliche*) mathematics and to its growth, and all problems relating to the situational logic of mathematical problem-solving.

I shall refer to the school of mathematical philosophy which tends to identify mathematics with its formal axiomatic abstraction (and the philosophy of mathematics with metamathematics) as the 'formalist' school. One of the clearest statements of the formalist position is to be found in Carnap [1937]. Carnap demands that (*a*) 'philosophy is to be replaced by the logic of science...', (*b*) 'the logic of science is nothing other than the logical syntax of the language of science...', (*c*) 'metamathematics is the syntax of mathematical language' (pp. xiii and 9). Or: philosophy of mathematics is to be replaced by metamathematics.

Formalism disconnects the history of mathematics from the philosophy of mathematics, since, according to the formalist concept of

---

[1] Church [1956], I, pp. 76–7. Also cf. Peano [1894], p. 49 and Russell and Whitehead [1910–13], I, p. 12. This is an integral part of the Euclidean programme as formulated in Pascal [1659]: cf. Lakatos [1962], p. 158.

mathematics, there is no history of mathematics proper. Any formalist would basically agree with Russell's 'romantically' put but seriously meant remark, according to which Boole's *Laws of Thought* (1854) was 'the first book ever written on mathematics'.[1] Formalism denies the status of mathematics to most of what has been commonly understood to be mathematics, and can say nothing about its growth. None of the 'creative' periods and hardly any of the 'critical' periods of mathematical theories would be admitted into the formalist heaven, where mathematical theories dwell like the seraphim, purged of all the impurities of earthly uncertainty. Formalists, though, usually leave open a small back door for fallen angels: if it turns out that for some 'mixtures of mathematics and something else' we can find formal systems 'which include them in a certain sense', then they too may be admitted (Curry [1951], pp. 56–7). On those terms Newton had to wait four centuries until Peano, Russell, and Quine helped him into heaven by formalising the Calculus. Dirac is more fortunate: Schwartz saved his soul during his lifetime. Perhaps we should mention here the paradoxical plight of the metamathematician: by formalist, or even by deductivist, standards, he is not an honest mathematician. Dieudonné talks about 'the absolute necessity imposed on any mathematician *who cares for intellectual integrity*' (my italics) to present his reasonings in axiomatic form ([1939], p. 225).

Under the present dominance of formalism, one is tempted to paraphrase Kant: the history of mathematics, lacking the guidance of philosophy, has become *blind*, while the philosophy of mathematics, turning its back on the most intriguing phenomena in the history of mathematics, has become *empty*.

'Formalism' is a bulwark of logical positivist philosophy. According to logical positivism, a statement is meaningful only if it is either 'tautological' or empirical. Since informal mathematics is neither 'tautological' nor empirical, it must be meaningless, sheer nonsense.[2]

---

[1] Russell [1901]. The essay was republished as chapter 5 of Russell's [1918], under the title 'Mathematics and the Metaphysicians'. In the 1953 Penguin edition the quotation can be found on p. 74. In the preface of his [1918] Russell says of the essay: 'Its tone is partly explained by the fact that the editor begged me to make the article "as romantic as possible".'

[2] According to Turquette, Gödelian sentences are meaningless ([1950], p. 129). Turquette argues against Copi who claims that since they are *a priori truths* but not analytic, they refute the analytic theory of *a priori* ([1949] and [1950]). Neither of them notices that the peculiar status of Gödelian sentences from this point of view is that these theorems are theorems of informal mathematics, and that in fact they are discussing the status of informal mathematics in a particular case.

The dogmas of logical positivism have been detrimental to the *history and philosophy of mathematics.*

The purpose of these essays is to approach some problems of the *methodology of mathematics.* I use the word 'methodology' in a sense akin to Pólya's and Bernays' 'heuristic'[1] and Popper's 'logic of discovery' or 'situational logic'.[2] The recent expropriation of the term 'methodology of mathematics' to serve as a synonym for 'metamathematics' has undoubtedly a formalist touch. It indicates that in formalist philosophy of mathematics there is no proper place for methodology qua logic of discovery.[3] According to formalists, mathematics is identical with formalised mathematics. But what can one

---

[1] Pólya [1945], especially p. 102, and also [1954], [1962*a*]; Bernays [1947], esp. p. 187.

[2] Popper [1934], then [1945], especially p. 90 (or the fourth edition [1962], p. 97); and also [1957], pp. 147 ff.

[3] One can illustrate this, e.g. by Tarski [1930*a*] and Tarski [1930*b*]. In the first paper Tarski uses the term 'deductive sciences' *explicitly* as a shorthand for 'formalised deductive sciences'. He says: 'Formalised deductive disciplines form the field of research of metamathematics roughly in the same sense in which spatial entities form the field of research in geometry.' This sensible formulation is given an intriguing imperialist twist in the second paper: 'The deductive disciplines constitute the subject-matter of the methodology of the deductive sciences in much the same sense in which spatial entities constitute the subject-matter of geometry and animals that of zoology. Naturally not all deductive disciplines are presented in a form suitable for objects of scientific investigation. Those, for example, are not suitable which do not rest on a definite logical basis, have no precise rules of inference, and the theorems of which are formulated in the usually ambiguous and inexact terms of colloquial language – in a word those which are not formalised. Metamathematical investigations are confined in consequence to the discussion of formalised deductive disciplines.' The innovation is that while the first formulation stated that the subject matter of metamathematics is the formalised deductive disciplines, the second formulation states that the subject-matter of metamathematics is confined to formalised deductive disciplines only because non-formalised deductive sciences are not suitable objects for scientific investigation at all. This implies that the pre-history of a formalised discipline cannot be the subject-matter of a scientific investigation – unlike the pre-history of a zoological species, which can be the subject-matter of a very scientific theory of evolution. Nobody will doubt that some problems about a mathematical theory can only be approached after it has been formalised, just as some problems about human beings (say concerning their anatomy) can only be approached after their death. But few will infer from this that human beings are 'suitable for scientific investigation' only when they are 'presented in "dead" form', and that biological investigations are confined in consequence to the discussion of dead human beings – although, I should not be surprised if some enthusiastic pupil of Vesalius in those glorious days of early anatomy, when the powerful new method of dissection emerged, had identified biology with the analysis of dead bodies.

In the preface of his [1941] Tarski enlarges on his negative attitude towards the possibility of any sort of methodology other than formal systems: 'A course in the methodology of empirical sciences...must be largely confined to evaluations and criticisms of tentative gropings and unsuccessful efforts.' The reason is that empirical sciences are unscientific: for Tarski defines a scientific theory 'as a system of asserted statements arranged according to certain rules' (ibid.).

discover in a formalised theory? Two sorts of things. *First*, one can discover the solution to problems which a suitably programmed Turing machine could solve in a finite time (such as: is a certain alleged proof a proof or not?). No mathematician is interested in following out the dreary mechanical 'method' prescribed by such decision procedures. *Secondly*, one can discover the solutions to problems (such as: is a certain formula in a non-decidable theory a theorem or not?), where one can be guided only by the 'method' of 'unregimented insight and good fortune'.

Now this bleak alternative between the rationalism of a machine and the irrationalism of blind guessing does not hold for live mathematics:[1] an investigation of *informal* mathematics will yield a rich situational logic for working mathematicians, a situational logic which is neither mechanical nor irrational, but which cannot be recognised and still less, stimulated, by the formalist philosophy.

The history of mathematics and the logic of mathematical discovery, i.e. the phylogenesis and the ontogenesis of mathematical thought,[2] cannot be developed without the criticism and ultimate rejection of formalism.

But formalist philosophy of mathematics has very deep roots. It is the latest link in the long chain of *dogmatist* philosophies of mathematics. For more than two thousand years there has been an argument between *dogmatists* and *sceptics*. The dogmatists hold that – by the power of our human intellect and/or senses – we can attain truth and know that we have attained it. The sceptics on the other hand either hold that we cannot attain the truth at all (unless with the help of mystical experience), or that we cannot know if we can attain it or that

---

[1] One of the most dangerous vagaries of formalist philosophy is the habit of (1) stating something – rightly – about formal systems; (2) then saying that this applies to 'mathematics ' – this is again right if we accept the identification of mathematics and formal systems; (3) subsequently, with a surreptitious shift in meaning, using the term 'mathematics' in the ordinary sense. So Quine says ([1951], p. 87) that 'this reflects the characteristic mathematical situation; the mathematician hits upon his proof by unregimented insight and good fortune, but afterwards other mathematicians can check his proof'. But often the checking of an *ordinary* (informal) proof is a very delicate enterprise, and to hit on a 'mistake' requires as much insight and luck as to hit on a proof: the discovery of 'mistakes' in informal proofs may sometimes take decades – if not centuries.

[2] Both H. Poincaré and G. Pólya propose to apply E. Haeckel's 'fundamental biogenetic law' about ontogeny recapitulating phylogeny to mental development, in particular to mathematical mental development. (Poincaré [1908], p. 135, and Pólya [1962b].) To quote Poincaré: 'Zoologists maintain that the embryonic development of an animal recapitulates in brief the whole history of its ancestors throughout geologic time. It seems it is the same in the development of minds...For this reason, the history of science should be our first guide' (C. B. Halsted's authorised translation, p. 437).

we have attained it. In this great debate, in which arguments are time and again brought up to date, mathematics has been the proud fortress of dogmatism. Whenever the mathematical dogmatism of the day got into a 'crisis', a new version once again provided genuine rigour and ultimate foundations, thereby restoring the image of authoritative, infallible, irrefutable mathematics, 'the only Science that it has pleased God hitherto to bestow on mankind' (Hobbes [1651], p. 15). Most sceptics resigned themselves to the impregnability of this stronghold of dogmatist epistemology.[1] A challenge is now overdue.

The core of this case-study will challenge mathematical formalism, but will not challenge directly the ultimate positions of mathematical dogmatism. Its modest aim is to elaborate the point that informal, quasi-empirical, mathematics does not grow through a monotonous increase of the number of indubitably established theorems but through the incessant improvement of guesses by speculation and criticism, by the logic of proofs and refutations. Since, however, metamathematics is a paradigm of informal, quasi-empirical mathematics just now in rapid growth, the essay, by implication, will also challenge modern mathematical dogmatism. The student of recent history of meta-mathematics will recognise the patterns described here in his own field.

The dialogue form should reflect the dialectic of the story; it is meant to contain a sort of *rationally reconstructed or 'distilled' history. The real history will chime in in the footnotes, most of which are to be taken, therefore, as an organic part of the essay.*

---

[1] For a discussion of the rôle of mathematics in the dogmatist-sceptic controversy, cf. my [1962].

I

1. A Problem and a Conjecture

The dialogue takes place in an imaginary classroom. The class gets interested in a PROBLEM: is there a relation between the number of vertices V, the number of edges E and the number of faces F of polyhedra – particularly of *regular polyhedra* – analogous to the trivial relation between the number of vertices and edges of *polygons*, namely, that there are as many edges as vertices: $V = E$? This latter relation enables us to classify *polygons* according to the number of edges (or vertices): triangles, quadrangles, pentagons, etc. An analogous relation would help to classify *polyhedra*.

After much trial and error they notice that for all regular polyhedra $V - E + F = 2$.[1] Somebody *guesses* that this may apply for any polyhedron whatsoever. Others try to falsify this *conjecture*, try to test it in many different ways – it holds good. The results *corroborate* the conjecture, and suggest that it could be *proved*. It is at this point – after the

---

[1] First noticed by Euler [1758a]. His original problem was the classification of polyhedra, the difficulty of which was pointed out in the editorial summary: 'While in plane geometry polygons (*figurae rectilineae*) could be classified very easily according to the number of their sides, which of course is always equal to the number of their angles, in stereometry the classification of polyhedra (*corpora hedris planis inclusa*) represents a much more difficult problem, since the number of faces alone is insufficient for this purpose.'

The key to Euler's result was just the invention of the concepts of *vertex* and *edge*: it was he who first pointed out that besides the number of faces the number of *points* and *lines* on the surface of the polyhedron determines its (topological) character. It is interesting that on the one hand he was eager to stress the novelty of his conceptual framework, and that he had to invent the term '*acies*' (edge) instead of the old '*latus*' (side), since *latus* was a polygonal concept while he wanted a polyhedral one, on the other hand he still retained the term '*angulus solidus*' (solid angle) for his point-like vertices. It has been recently generally accepted that the priority of the result goes to Descartes. The ground for this claim is a manuscript of Descartes [*c.* 1639] copied by Leibniz in Paris from the original in 1675–6, and rediscovered and published by Foucher de Careil in 1860. The priority should not be granted to Descartes without a minor qualification. It is true that Descartes states that the number of plane angles equals $2\phi + 2\alpha - 4$ where by ϕ he means the number of faces and by α the number of solid angles. It is also true that he states that there are twice as many plane angles as edges (*latera*). The conjunction of these two statements of course yields the Euler formula. But Descartes did not see the point of doing so, since he still thought in terms of angles (plane and solid) and faces, and did not make a conscious revolutionary change to the concepts of 0-dimensional vertices, 1-dimensional edges and 2-dimensional faces as a necessary and sufficient basis for the full topological characterisation of polyhedra.

6

stages *problem* and *conjecture* – that we enter the classroom.[1] The teacher is just going to offer a *proof*.

2. A Proof

TEACHER: In our last lesson we arrived at a conjecture concerning polyhedra, namely, that for all polyhedra $V - E + F = 2$, where V is the number of vertices, E the number of edges and F the number of faces. We tested it by various methods. But we haven't yet proved it. Has anybody found a proof?

PUPIL SIGMA: 'I for one have to admit that I have not yet been able to devise a strict proof of this theorem... As however the truth of it has been established in so many cases, there can be no doubt that it holds good for any solid. Thus the proposition seems to be satisfactorily demonstrated.'[2] But if you have a proof, please do present it.

TEACHER: In fact I have one. It consists of the following thought-experiment. *Step 1:* Let us imagine the polyhedron to be hollow, with a surface made of thin rubber. If we cut out one of the faces, we can stretch the remaining surface flat on the blackboard, without tearing it. The faces and edges will be deformed, the edges may become curved, but V and E will not alter, so that if and only if $V - E + F = 2$ for the original polyhedron, $V - E + F = 1$ for this flat network – remember that we have removed one face. (Fig. 1 shows the flat network for the case of a cube.) *Step 2:* Now we triangulate our map – it does indeed look like a geographical map. We draw (possibly curvilinear) diagonals in those (possibly curvilinear) polygons which are not already (possibly curvilinear) triangles. By drawing each diagonal we increase both E and F by one, so that the total $V - E + F$ will not be altered (fig. 2). *Step 3:* From the triangulated network we now remove the triangles one by one. To remove a triangle we either remove an edge – upon which one face and one edge disappear (fig. 3(a)), or we remove two edges and a vertex – upon which one face, two edges and one vertex disappear (fig. 3(b)). Thus if $V - E + F = 1$ before a triangle is removed,

---

[1] Euler tested the conjecture quite thoroughly for consequences. He checked it for prisms, pyramids and so on. He could have added that the proposition that there are only five regular bodies is also a consequence of the conjecture. Another suspected consequence is the hitherto corroborated proposition that four colours are sufficient to colour a map.

The phase of *conjecturing* and *testing* in the case of $V - E + F = 2$ is discussed in Pólya ([1954], vol. 1, the first five sections of the third chapter, pp. 35–41). Pólya stopped here, and does not deal with the phase of *proving* – though of course he points out the need for a heuristic of 'problems to prove' ([1945], p. 144). Our discussion starts where Pólya stops.

[2] Euler ([1758a], p. 119 and p. 124). But later ([1758b]) he proposed a proof.

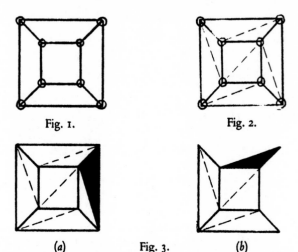

Fig. 1.

Fig. 2.

(a)

Fig. 3.

(b)

it remains so after the triangle is removed. At the end of this procedure we get a single triangle. For this $V - E + F = 1$ holds true. Thus we have proved our conjecture.[1]

PUPIL DELTA: You should now call it a *theorem*. There is nothing conjectural about it any more.[2]

PUPIL ALPHA: I wonder. I see that this experiment can be performed for a cube or for a tetrahedron, but how am I to know that it can be performed for *any* polyhedron? For instance, are you sure, Sir, that *any polyhedron, after having a face removed, can be stretched flat on the blackboard?* I am dubious about your first step.

PUPIL BETA: Are you sure that *in triangulating the map one will always get a new face for any new edge?* I am dubious about your second step.

PUPIL GAMMA: Are you sure that *there are only two alternatives – the disappearance of one edge or else of two edges and a vertex – when one drops the triangles one by one?* Are you even sure that *one is left with a single triangle at the end of this process?* I am dubious about your third step.[3]

TEACHER: Of course I am not sure.

[1] This proof-idea stems from Cauchy [1813a].

[2] Delta's view that this proof has established the 'theorem' beyond doubt was shared by many mathematicians in the nineteenth century, e.g. Crelle [1826–7], 2, pp. 668–71, Matthiessen [1863], p. 449, Jonquières [1890a] and [1890b]. To quote a characteristic passage: 'After Cauchy's proof, it became absolutely indubitable that the elegant relation $V + F = E + 2$ applies to all sorts of polyhedra, just as Euler stated in 1752. In 1811 all indecision should have disappeared.' Jonquières [1890a], pp. 111–12.

[3] The class is a rather advanced one. To Cauchy, Poinsot, and to many other excellent mathematicians of the nineteenth century these questions did not occur.

ALPHA: But then we are worse off than before! Instead of one conjecture we now have at least three! And this you call a 'proof'!

TEACHER: I admit that the traditional name 'proof' for this thought-experiment may rightly be considered a bit misleading. I do not think that it establishes the truth of the conjecture.

DELTA: What does it do then? What do you think a mathematical proof proves?

TEACHER: This is a subtle question which we shall try to answer later. Till then I propose to retain the time-honoured technical term 'proof' for a *thought-experiment – or 'quasi-experiment' – which suggests a decomposition of the original conjecture into subconjectures or lemmas,* thus *embedding it* in a possibly quite distant body of knowledge. Our 'proof', for instance, has embedded the original conjecture – about crystals, or, say, solids – in the theory of rubber sheets. Descartes or Euler, the fathers of the original conjecture, certainly did not even dream of this.[1]

---

[1] Thought-experiment (*deiknymi*) was the most ancient pattern of mathematical proof. It prevailed in pre-Euclidean Greek mathematics (cf. Á. Szabó [1958]).

That conjectures (or theorems) precede proofs in the heuristic order was a commonplace for ancient mathematicians. This followed from the heuristic precedence of '*analysis*' over '*synthesis*'. (For an excellent discussion see Robinson [1936].) According to Proclos, '...it is...necessary to know beforehand what is sought' (Heath [1925], 1, p. 129). 'They said that a theorem is that which is proposed with a view to the demonstration of the very thing proposed' – says Pappus (ibid. 1, p. 10). The Greeks did not think much of propositions which they happened to hit upon in the deductive direction without having previously guessed them. They called them *porisms*, corollaries, incidental results springing from the proof of a theorem or the solution of a problem, results not directly sought but appearing, as it were, by chance, without any additional labour, and constituting, as Proclus says, a sort of windfall (*ermaion*) or bonus (*kerdos*) (ibid. 1, p. 278). We read in the editorial summary to Euler [1756–7] that arithmetical theorems 'were discovered long before their truth has been confirmed by rigid demonstrations'. Both the Editor and Euler use for this process of discovery the modern term '*induction*' instead of the ancient '*analysis*' (ibid.). The heuristic precedence of the result over the argument, of the theorem over the proof, has deep roots in mathematical folklore. Let us quote some variations on a familiar theme: Chrysippus is said to have written to Cleanthes: 'Just send me the theorems, then I shall find the proofs' (cf. Diogenes Laertius [c. 200], VII. 179). Gauss is said to have complained: 'I have had my results for a long time; but I do not yet know how I am to arrive at them' (cf. Arber [1945], p. 47), and Riemann: 'If only I had the theorems! Then I should find the proofs easily enough.' (Cf. Hölder [1924], p. 487.) Pólya stresses: 'You have to guess a mathematical theorem before you prove it' ([1954], vol. 1, p. vi).

The term '*quasi-experiment*' is from the above-mentioned editorial summary to Euler [1753]. According to the Editor: 'As we must refer the numbers to the pure intellect alone, we can hardly understand how observations and *quasi-experiments* can be of use in investigating the nature of the numbers. Yet, in fact, as I shall show here with very good reasons, the properties of the numbers known today have been mostly discovered by observation...' (Pólya's translation; in his [1954], 1, p. 3 he mistakenly attributes the quotation to Euler).

3. Criticism of the Proof by Counterexamples which are Local but not Global

TEACHER: This decomposition of the conjecture suggested by the proof opens new vistas for testing. The decomposition deploys the conjecture on a wider front, so that our criticism has more targets. We now have at least three opportunities for counterexamples instead of one!

GAMMA: I have already expressed my dislike of your third lemma (viz. that in removing triangles from the network which resulted from the stretching and subsequent triangulation, we have only two possibilities: either we remove an edge or we remove two edges and a vertex). I suspect that other patterns may emerge when removing a triangle.

TEACHER: Suspicion is not criticism.

GAMMA: Then is a *counterexample* criticism?

TEACHER: Certainly. Conjectures ignore dislike and suspicion, but they cannot ignore counterexamples.

THETA (*aside*): Conjectures are obviously very different from those who represent them.

GAMMA: I propose a trivial counterexample. Take the triangular network which results from performing the first two operations on a cube (fig. 2). Now if I remove a triangle from the *inside* of this network, as one might take a piece out of a jigsaw puzzle, I remove one triangle without removing a single edge or vertex. So the third lemma is false – and not only in the case of the cube, but for *all* polyhedra except the tetrahedron, in the flat network of which all the triangles are boundary triangles. Your proof thus proves the Euler theorem for the tetrahedron. But we already *knew* that $V - E + F = 2$ for the tetrahedron, so why prove it?

TEACHER: You are right. But notice that the cube which is a counterexample to the third lemma is not also a counterexample to the main conjecture, since for the cube $V - E + F = 2$. You have shown the poverty of the argument – the proof – but not the falsity of our conjecture.

ALPHA: Will you scrap your proof then?

TEACHER: No. Criticism is not necessarily destruction. I shall improve my proof so that it will stand up to the criticism.

GAMMA: How?

TEACHER: Before showing how, let me introduce the following terminology. I shall call a '*local counterexample*' an example which

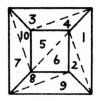

Fig. 4.

refutes a lemma (without necessarily refuting the main conjecture), and I shall call a '*global counterexample*' an example which refutes the main conjecture itself. Thus your counterexample is local but not global. A local, but not global, counterexample is a criticism of the proof, but not of the conjecture.

GAMMA: So, the conjecture may be true, but your proof does not prove it.

TEACHER: But I can easily elaborate, and *improve the proof*, by replacing the false lemma by a slightly modified one, which your counterexample will not refute. I no longer contend that *the removal of any triangle follows one of the two patterns mentioned*, but merely that *at each stage of the removing operation the removal of any boundary triangle follows one of these patterns*. Coming back to my thought-experiment, all that I have to do is to insert a single word in my third step, to wit, that 'from the triangulated network we now remove the *boundary* triangles one by one'. You will agree that it only needed a trifling observation to put the proof right.[1]

GAMMA: I do not think your observation was so trifling; in fact it was quite ingenious. To make this clear I shall show that it is false. Take the flat network of the cube again and remove eight of the ten triangles in the order given in fig. 4. At the removal of the eighth triangle, which is certainly by then a boundary triangle, we removed two edges and no vertex – this changes $V - E + F$ by 1. And we are left with the two disconnected triangles 9 and 10.

TEACHER: Well, I might save face by saying that I meant by a boundary triangle a triangle whose removal does not disconnect the network. But intellectual honesty prevents me from making surreptitious changes in my position by sentences starting with 'I meant ...' so I admit that now I must *replace* the second version of the

---

[1] Lhuilier, when correcting in a similar way a proof of Euler, says that he made only a 'trifling observation' ([1812–13a], p. 179). Euler himself, however, gave the proof up, since he noticed the trouble but could not make that 'trifling observation'.

triangle-removing operation with a third version: that we remove the triangles one by one in such a way that $V-E+F$ does not alter.

KAPPA: I generously agree that the lemma corresponding to this operation is true: namely, that if we remove the triangles one by one in such a way that $V-E+F$ does not alter, then $V-E+F$ does not alter.

TEACHER: No. The lemma is that *the triangles in our network can be so numbered that in removing them in the right order $V-E+F$ will not alter till we reach the last triangle.*

KAPPA: But how should one construct this right order, if it exists at all?[1] Your original thought-experiment gave the instruction: remove the triangles in any order. Your modified thought-experiment gave the instruction: remove boundary triangles in any order. Now you say we should follow a definite order, but you do not say which and whether that order exists at all. Thus the thought-experiment breaks down. You improved the proof-analysis, i.e. the list of lemmas; but the thought-experiment which you called 'the proof' has disappeared.

RHO: Only the third step has disappeared.

KAPPA: Moreover, did you *improve* the lemma? Your first two simple versions at least looked trivially true before they were refuted; your lengthy, patched up version does not even look plausible. Can you really believe that it will escape refutation?

TEACHER: 'Plausible' or even 'trivially true' propositions are usually soon refuted: sophisticated, implausible conjectures, matured in criticism, might hit on the truth.

OMEGA: And what happens if even your 'sophisticated conjectures' are falsified and if this time you cannot replace them by unfalsified ones? Or, if you do *not* succeed in improving the argument further by local patching? You have succeeded in getting over a local counter-example which was not global by replacing the refuted lemma. What if you do not succeed next time?

TEACHER: Good question – it will be put on the agenda for tomorrow.

---

[1] Cauchy thought that the instruction to find at each stage a triangle which can be removed either by removing two edges and a vertex or one edge can be trivially carried out for any polyhedron ([1813a], p. 79). This is connected with his inability to imagine a polyhedron that is not homeomorphic with the sphere.

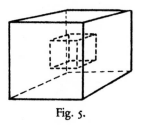

Fig. 5.

4. Criticism of the Conjecture by Global Counterexamples

ALPHA: I have a counterexample which will falsify your first lemma – but this will also be a counterexample to the main conjecture, i.e. this will be a global counterexample as well.

TEACHER: Indeed! Interesting. Let us see.

ALPHA: Imagine a solid bounded by a pair of nested cubes – a pair of cubes, one of which is inside, but does not touch the other (fig. 5). This hollow cube falsifies your first lemma, because on removing a face from the inner cube, the polyhedron will not be stretchable on to a plane. Nor will it help to remove a face from the outer cube instead. Besides, for each cube $V - E + F = 2$, so that for the hollow cube $V - E + F = 4$.

TEACHER: Good show. Let us call it *Counterexample 1*.[1] Now what?

(a) Rejection of the conjecture. The method of surrender

GAMMA: Sir, your composure baffles me. A single counterexample refutes a conjecture as effectively as ten. The conjecture and its proof have completely misfired. Hands up! You have to surrender. Scrap the false conjecture, forget about it and try a radically new approach.

TEACHER: I agree with you that the *conjecture* has received a severe criticism by Alpha's counterexample. But it is untrue that the *proof* has 'completely misfired'. If, for the time being, you agree to my earlier proposal to use the word 'proof' for a 'thought-experiment' which leads to decomposition of the original conjecture into•

---

[1] This *Counterexample 1* was first noticed by Lhuilier ([1812–13a], p. 194). But Gergonne the Editor, added (p. 186) that he himself noticed this long before Lhuilier's paper. Not so Cauchy, who published his proof just a year before. And this counterexample was to be rediscovered twenty years later by Hessel ([1832], p. 16). Both Lhuilier and Hessel were led to their discovery by mineralogical collections in which they noticed some double crystals, where the inner crystal is not translucent, but the outer is. Lhuilier acknowledges the stimulus of the crystal collection of his friend Professor Pictet ([1812–13a], p. 188). Hessel refers to lead sulphide cubes enclosed in translucent calcium fluoride crystals ([1832], p. 16).

subconjectures', instead of using it in the sense of a 'guarantee of certain truth', you need not draw this conclusion. My proof certainly proved Euler's conjecture in the first sense, but not necessarily in the second. You are interested only in proofs which 'prove' what they have set out to prove. I am interested in proofs even if they do not accomplish their intended task. Columbus did not reach India but he discovered something quite interesting.

ALPHA: So according to your philosophy – while a local counterexample (if it is not global at the same time) is a criticism of the proof, but not of the conjecture – a global counterexample is a criticism of the conjecture, but not necessarily of the proof. You agree to surrender as regards the conjecture, but you defend the proof. But if the conjecture is false, what on earth does the proof prove?

GAMMA: Your analogy with Columbus breaks down. Accepting a global counterexample must mean total surrender.

(b) Rejection of the counterexample. The method of monster-barring

DELTA: But why accept the counterexample? We proved our conjecture – now it is a theorem. I admit that it clashes with this so-called 'counterexample'. One of them has to give way. But why should the theorem give way, when it has been proved? It is the 'criticism' that should retreat. It is fake criticism. This pair of nested cubes is not a polyhedron at all. It is a *monster*, a pathological case, not a counterexample.

GAMMA: Why not? *A polyhedron is a solid whose surface consists of polygonal faces.* And my counterexample is a solid bounded by polygonal faces.

TEACHER: Let us call this definition *Def. 1*.[1]

DELTA: Your definition is incorrect. A polyhedron must be a *surface*: it has faces, edges, vertices, it can be deformed, stretched out on a blackboard, and has nothing to do with the concept of 'solid'. *A polyhedron is a surface consisting of a system of polygons.*

TEACHER: Call this *Def. 2*.[2]

DELTA: So really you showed us *two* polyhedra – *two* surfaces, one

---

[1] *Definition 1* occurs first in the eighteenth century; e.g.: 'One gives the name *polyhedral solid*, or simply *polyhedron*, to any solid bounded by planes or plane faces' (Legendre [1809], p. 160). A similar definition is given by Euler ([1758a]). Euclid, while defining cube, octahedron, pyramid, prism, does not define the general term polyhedron, but occasionally uses it (e.g. Book XII, Second Problem, Prop. 17).

[2] We find *Definition 2* implicitly in one of Jonquières' papers read to the French Academy against those who meant to refute Euler's theorem. These papers are a thesaurus of monster-barring techniques. He thunders against Lhuilier's monstrous pair of nested

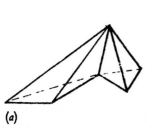

(a)   Fig. 6.   (b)

completely inside the other. A woman with a child in her womb is not a counterexample to the thesis that human beings have one head.

ALPHA: So! My counterexample has bred a new concept of polyhedron. Or do you dare to assert that by polyhedron you *always* meant a surface?

TEACHER: For the moment let us accept Delta's *Def. 2*. Can you refute our conjecture now if by polyhedron we mean a surface?

ALPHA: Certainly. Take two tetrahedra which have an edge in common (fig. 6(a)). Or, take two tetrahedra which have a vertex in common (fig. 6(b)). Both these twins are connected, both constitute one single surface. And, you may check that for both $V - E + F = 3$.

TEACHER: *Counterexamples 2a and 2b.*[1]

DELTA: I admire your perverted imagination, but of course I did not mean that *any* system of polygons is a polyhedron. By polyhedron I meant *a system of polygons arranged in such a way that (1) exactly two polygons meet at every edge and (2) it is possible to get from the inside of any polygon to the inside of any other polygon by a route which never crosses any edge at a vertex.* Your first twins will be excluded by the first criterion in my definition, your second twins by the second criterion.

TEACHER: *Def. 3.*[2]

cubes: 'Such a system is not really a polyhedron but a pair of distinct polyhedra, each independent of the other...A polyhedron, at least from the classical point of view, deserves the name only if, before all else, a point can move continuously over its entire surface; here this is not the case...This first exception of Lhuilier can therefore be discarded' ([1890b], p. 170). This definition – as opposed to Definition 1 – goes down very well with analytical topologists who are not interested at all in the theory of polyhedra as such but only as a handmaiden for the theory of surfaces.

[1] *Counterexamples 2a* and *2b* were missed by Lhuilier and first discovered only by Hessel ([1832], p. 13).

[2] *Definition 3* first turns up to keep out twintetrahedra in Möbius ([1865], p. 32). We find his cumbersome definition reproduced in some modern textbooks in the usual authoritarian 'take it or leave it' way; the story of its monster-barring background – that would at least explain it – is not told (e.g. Hilbert and Cohn-Vossen [1956], p. 290).

ALPHA: I admire your perverted ingenuity in inventing one definition after another as barricades against the falsification of your pet ideas. Why don't you just define a polyhedron as a system of polygons for which the equation $V - E + F = 2$ holds? This Perfect Definition...

KAPPA: *Def. P.*[1]

ALPHA: ...would settle the dispute for ever. There would be no need to investigate the subject any further.

DELTA: But there isn't a theorem in the world which couldn't be falsified by monsters.

TEACHER: I am sorry to interrupt you. As we have seen, refutation by counterexamples depends on the meaning of the terms in question. If a counterexample is to be an objective criticism, we have to agree on the meaning of our terms. We *may* achieve such an agreement by defining the term where communication broke down. I, for one, didn't define 'polyhedron'. I assumed *familiarity* with the concept, i.e. the ability to distinguish a thing which is a polyhedron from a thing which is not a polyhedron – what some logicians call knowing the extension of the concept of polyhedron. It turned out that the extension of the concept wasn't at all obvious: *definitions are frequently proposed and argued about when counterexamples emerge.* I suggest that we now consider the rival definitions together, and leave until later the discussion of the differences in the results which will follow from choosing different definitions. Can anybody offer something which even the most restrictive definition would allow as a real counterexample?

KAPPA: Including *Def. P*?

TEACHER: Excluding *Def. P*.

GAMMA: I can. Look at this *Counterexample 3*: a star-polyhedron – I shall call it an *urchin* (fig. 7). This consists of 12 star-pentagons (fig. 8). It has 12 vertices, 30 edges, and 12 pentagonal faces – you may check it if you like by counting. Thus the Descartes–Euler thesis is not true at all, since for this polyhedron $V - E + F = -6$.[2]

---

[1] *Definition P*, according to which Eulerianness would be a definitional characteristic of polyhedra, was in fact suggested by R. Baltzer: 'Ordinary polyhedra are occasionally (following Hessel) called Eulerian polyhedra. It would be more appropriate to find a special name for non-genuine (*uneigentliche*) polyhedra' ([1862], vol. 2, p. 207). The reference to Hessel is unfair: Hessel used the term 'Eulerian' simply as an abbreviation for polyhedra for which Euler's relation holds in contradistinction to the non-Eulerian ones ([1832], p. 19). For *Def. P* see also the Schläfli quotation in footnote 2 below.

[2] The 'urchin' was first discussed by Kepler in his cosmological theory ([1619], Lib. II, XIX and XXVI, on p. 72 and pp. 82–3 and Lib. V, *Cap.* I, p. 293, *Cap.* III, p. 299 and *Cap.* IX, XLVII). The name 'urchin' is Kepler's ('*cui nomen Echino feci*'). Fig. 7 is copied

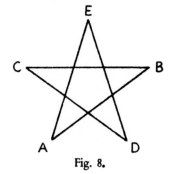

Fig. 7. Kepler's star-polyhedron, each face shaded in a different way to show which triangles belong to the same pentagonal face.

Fig. 8.

DELTA: Why do you think that your 'urchin' is a polyhedron?

GAMMA: Do you not see? This is a polyhedron, whose faces are the twelve star-pentagons. It satisfies your last definition: it is 'a system of polygons arranged in such a way that (1) exactly two polygons meet at every edge, and (2) it is possible to get from every polygon to every other polygon without ever crossing a vertex of the polyhedron'.

DELTA: But then you do not even know what a polygon is! A star-pentagon is certainly not a polygon! *A polygon is a system of edges arranged in such a way that (1) exactly two edges meet at every vertex, and (2) the edges have no points in common except the vertices.*

TEACHER: Let us call this *Def. 4.*

GAMMA: I don't see why you include the second clause. The correct definition of the polygon should contain the first clause only.

TEACHER: *Def. 4'.*

GAMMA: The second clause has nothing to do with the essence of a polygon. Look: if I lift an edge a little, the star-pentagon is already a polygon even in your sense. You imagine a polygon to be drawn in chalk on the blackboard, but you should imagine it as a wooden structure: then it is clear that what you think to be a point in common is not really one point, but two different points lying one above the

from his book (p. 79) which contains also another picture on p. 293. Poinsot independently rediscovered it, and it was he who pointed out that the Euler formula did not apply to it ([1810], p. 48). The now standard term 'small stellated dodecahedron' is Cayley's ([1859], p. 125). Schläfli admitted star-polyhedra in general, but nevertheless rejected our small stellated dodecahedron as a monster. According to him 'this is not a genuine polyhedron, for it does not satisfy the condition $V - E + F = 2$' ([1852], § 34).

other. You are misled by your embedding the polygon in a plane – you should let its limbs stretch out in space![1]

DELTA: Would you mind telling me what is the *area* of a star-pentagon? Or would you say that some polygons have no area?

GAMMA: Was it not you yourself who said that a polyhedron has nothing to do with the idea of solidity? Why now suggest that the idea of polygon should be linked with the idea of area? We agreed that a polyhedron is a closed surface with edges and vertices – then why not agree that a polygon is simply a closed curve with vertices? But if you stick to your idea I am willing to define the area of a star-polygon.[2]

TEACHER: Let us leave this dispute for a moment, and proceed as before. Consider the last two definitions together – *Def. 4* and *Def. 4'*.

[1] The dispute whether polygon should be defined so as to include star-polygons or not (*Def. 4* or *Def. 4'*) is a very old one. The argument put forward in our dialogue – that star-polygons become ordinary polygons when embedded in a space of higher dimensions – is a modern topological argument, but one can put forward many others. Thus Poinsot defending his star-polyhedra argued for the admission of star-polygons with arguments taken from analytical geometry: '...all these distinctions (between "ordinary" and "star"-polygons) are more apparent than real, and they completely disappear in the analytical treatment, in which the various species of polygons are quite inseparable. To the edge of a regular polygon there corresponds an equation with real roots, which simultaneously yields the edges of all the regular polygons of the same order. Thus it is not possible to obtain the edges of a regular inscribed heptagon, without at the same time finding edges of heptagons of the second and third species. Conversely, given the edge of a regular heptagon, one may determine the radius of a circle in which it can be inscribed, but in so doing, one will find three different circles corresponding to the three species of heptagon which may be constructed on the given edge; similarly for other polygons. Thus we are justified in giving the name "polygon" to these new starred figures' ([1810], p. 26). Schröder uses the Hankelian argument: 'The extension to rational fractions of the power concept originally associated only with the integers has been very fruitful in Algebra; this suggests that we try to do the same thing in geometry whenever the opportunity presents itself...' ([1862], p. 56). Then he shows that we may find a geometrical interpretation for the concept of p/q-sided polygons in the star-polygons.

[2] Gamma's claim that he can define the area for star-polygons is not a bluff. Some of those who defended the wider concept of polygon solved the problem by putting forward a wider concept of the area of polygon. There is an especially obvious way to do this in the case of regular star-polygons. We may take the area of a polygon as the sum of the areas of the isosceles triangles which join the centre of the inscribed or circumscribed circle to the sides. In this case, of course, some 'portions' of the star-polygon will count more than once. In the case of irregular polygons where we have not got any one distinguished point, we may still take any point as origin and treat negatively oriented triangles as having negative areas (Meister [1771], p. 179). It turns out – and this can certainly be expected from an 'area' – that the area thus defined will not depend on the choice of the origin (Möbius [1827], p. 218). Of course there is liable to be a dispute with those who think that one is not justified in calling the number yielded by this calculation an 'area'; though the defenders of the Meister–Möbius definition called it 'the right definition' which 'alone is scientifically justified' (R. Haussner's notes [1906], pp. 114–15). Essentialism has been a permanent feature of definitional quarrels.

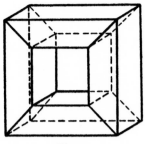

Fig. 9.

Can anyone give a counterexample to our conjecture that will comply with *both* definitions of polygons?

ALPHA: Here is one. Consider a *picture-frame* like this (fig. 9). This is a polyhedron according to any of the definitions hitherto proposed. Nonetheless you will find, on counting the vertices, edges and faces, that $V - E + F = 0$.

TEACHER: *Counterexample 4.*[1]

BETA: So that's the end of our conjecture. It really is a pity, since it held good for so many cases. But it seems that we have just wasted our time.

ALPHA: Delta, I am flabbergasted. You say nothing? Can't you define this new counterexample out of existence? I thought there was no hypothesis in the world which you could not save from falsification with a suitable linguistic trick. Are you giving up now? Do you agree at last that there exist non-Eulerian polyhedra? Incredible!

DELTA: You should really find a more appropriate name for your non-Eulerian pests and not mislead us all by calling them 'polyhedra'. But I am gradually losing interest in your monsters. I turn in disgust from your lamentable 'polyhedra', for which Euler's beautiful theorem doesn't hold.[2] I look for order and harmony in mathematics, but you only propagate anarchy and chaos.[3] Our attitudes are irreconcilable.

---

[1] We find *Counterexample 4* too in Lhuilier's classical [1812–13a], on p. 185 – Gergonne again added that he knew it. But Grunert did not know it fourteen years later ([1827]) nor did Poinsot forty-five years later ([1858], p. 67).

[2] This is paraphrased from a letter of Hermite's written to Stieltjes: 'I turn aside with a shudder of horror from this lamentable plague of functions which have no derivatives' ([1893]).

[3] 'Researches dealing with...functions violating laws which one hoped were universal, were regarded almost as the propagation of anarchy and chaos where past generations had sought order and harmony' (Saks [1933], Preface). Saks refers here to the fierce

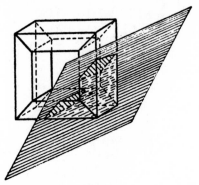

Fig. 10.

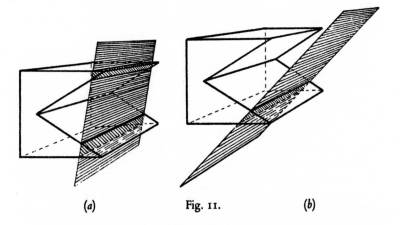

(a)   Fig. 11.   (b)

ALPHA: You are a real old-fashioned Tory! You blame the wickedness of anarchists for the spoiling of your 'order' and 'harmony', and you 'solve' the difficulties by verbal recommendations.

TEACHER: Let us hear the latest rescue-definition.

ALPHA: You mean the latest linguistic trick, the latest contraction of the concept of 'polyhedron'! Delta dissolves real problems, instead of solving them.

battles between monster-barrers (like Hermite!) and refutationists that characterised in the last decades of the nineteenth century (and indeed in the beginning of the twentieth) the development of modern real function theory, 'the branch of mathematics which deals with counterexamples' (Munroe [1953], Preface). The similarly fierce battle that raged later between the opponents and protagonists of modern mathematical logic and set-theory was a direct continuation of this. See also footnotes 2, p. 22, and 1, p. 23.

DELTA: I do not *contract* concepts. It is you who *expand* them. For instance, this picture-frame is not a genuine polyhedron at all.

ALPHA: Why?

DELTA: Take an arbitrary point in the 'tunnel' – the space bounded by the frame. Lay a plane through this point. You will find that any such plane has always *two* different cross-sections with the picture-frame, making two distinct, completely disconnected polygons! (fig. 10).

ALPHA: So what?

DELTA: *In the case of a genuine polyhedron, through any arbitrary point in space there will be at least one plane whose cross-section with the polyhedron will consist of one single polygon.* In the case of convex polyhedra *all* planes will comply with this requirement, wherever we take the point. In the case of *ordinary* concave polyhedra some planes will have more intersections, but there will always be some that have only one (fig. 11, (*a*) and (*b*)). In the case of this picture-frame, if we take the point in the tunnel, all the planes will have two cross-sections. How then can you call this a polyhedron?

TEACHER: This looks like another definition, this time an *implicit* one. Call it *Def. 5*.[1]

ALPHA: A series of counterexamples, a matching series of definitions, definitions that are alleged to contain nothing new, but to be merely new revelations of the richness of that one old concept, which seems to have as many 'hidden' clauses as there are counterexamples. *For all polyhedra* $V - E + F = 2$ seems unshakable, an old and 'eternal' truth. It is strange to think that once upon a time it was a wonderful guess, full of challenge and excitement. Now, because of your weird shifts of meaning, it has turned into a poor convention, a despicable piece of dogma. (*He leaves the classroom.*)

DELTA: I cannot understand how an able man like Alpha can waste his talent on mere heckling. He seems engrossed in the production of monstrosities. But monstrosities never foster growth, either in the world of nature or in the world of thought. Evolution always follows an harmonious and orderly pattern.

---

[1] *Definition 5* was put forward by the indefatigable monster-barrer E. de Jonquières to get Lhuilier's polyhedron with a tunnel (picture-frame) out of the way: 'Neither is this polyhedral complex a true polyhedron in the ordinary sense of the word, for if one takes any plane through an arbitrary point inside one of the tunnels which pass right through the solid, the resulting cross-section will be composed of two distinct polygons completely unconnected with each other; this can occur in an ordinary polyhedron for *certain* positions of the intersecting plane, namely in the case of some concave polyhedra, but not for all of them' ([1890*b*], pp. 170–1). One wonders whether de Jonquières has noticed that his *Def. 5* excludes also some concave spheroid polyhedra.

GAMMA: Geneticists can easily refute that. Have you not heard that mutations producing monstrosities play a considerable role in macro-evolution? They call such monstrous mutants 'hopeful monsters'. It seems to me that Alpha's counterexamples, though monsters, are 'hopeful monsters'.[1]

DELTA: Anyway, Alpha has given up the struggle. No more monsters now.

GAMMA: I have a new one. It complies with all the restrictions in Defs. 1, 2, 3, 4, and 5, but $V - E + F = 1$. This *Counterexample 5* is a simple cylinder. It has 3 faces (the top, the bottom and the jacket), 2 edges (two circles) and no vertices. It is a polyhedron according to your definition: (1) exactly two polygons at every edge and (2) it is possible to get from the inside of any polygon to the inside of any other polygon by a route which never crosses any edge at a vertex. And you have to accept the faces as genuine polygons, as they comply with your requirements: (1) exactly two edges meet at every vertex and (2) the edges have no points in common except the vertices.

DELTA: Alpha stretched concepts, but you tear them! Your 'edges' are not edges! *An edge has two vertices!*

TEACHER: *Def. 6?*

GAMMA: But why deny the status of 'edge' to edges with one or possibly zero vertices? You used to contract concepts, but now you mutilate them so that scarcely anything remains!

DELTA: But don't you see the futility of these so-called refutations? 'Hitherto, when a new polyhedron was invented, it was for some practical end; today they are invented expressly to put at fault the reasonings of our fathers, and one never will get from them anything more than that. Our subject is turned into a teratological museum where decent ordinary polyhedra may be happy if they can retain a very small corner.'[2]

---

[1] 'We must not forget that what appears to-day as a monster will be to-morrow the origin of a line of special adaptations...I further emphasized the importance of rare but extremely consequential mutations affecting rates of decisive embryonic processes which might give rise to what one might term hopeful monsters, monsters which would start a new evolutionary line if fitting into some empty environmental niche.' (Goldschmidt [1933], pp. 544 and 547). My attention was drawn to this paper by Karl Popper.

[2] Paraphrased from Poincaré ([1908], pp. 131–2). The original full text is this: 'Logic sometimes makes monsters. Since half a century we have seen arise a crowd of bizarre functions which seem to try to resemble as little as possible the honest functions which serve some purpose. No longer continuity, or perhaps continuity, but no derivatives, etc. Nay more, from the logical point of view, it is these strange functions which are the most general, those one meets without seeking no longer appear except as particular case. There remains for them only a very small corner.

GAMMA: I think that if we want to learn about anything really deep, we have to study it not in its 'normal', regular, usual form, but in its critical state, in fever, in passion. If you want to know the normal healthy body, study it when it is abnormal, when it is ill. If you want to know functions, study their singularities. If you want to know ordinary polyhedra, study their lunatic fringe. This is how one can carry mathematical analysis into the very heart of the subject.[1] But even if you were basically right, don't you see the futility of your *ad hoc* method? If you want to draw a borderline between counterexamples and monsters, you cannot do it in fits and starts.

TEACHER: I think we should refuse to accept Delta's strategy for dealing with global counterexamples, although we should congratulate him on his skilful execution of it. We could aptly label his method *the method of monster-barring*. Using this method one can eliminate any counterexample to the original conjecture by a sometimes deft but always *ad hoc* redefinition of the polyhedron, of its defining terms, or of the defining terms of its defining terms. We should somehow treat counterexamples with more respect, and not stubbornly exorcise them by dubbing them monsters. Delta's main mistake is perhaps his dogmatist bias in the interpretation of mathematical proof: he thinks that a proof necessarily proves what it has set out to prove. My interpretation of proof will allow for a *false* conjecture to be 'proved', i.e. to be decomposed into subconjectures. If the conjecture is false, I certainly expect at least one of the subconjectures to be false. But the decomposition might still be interesting! I am not perturbed at finding a counterexample to a 'proved' conjecture; I am even willing to set out to 'prove' a false conjecture!

THETA: I don't follow you.

KAPPA: He just follows the New Testament: 'Prove all things; hold fast that which is good' (1 Thessalonians 5: 21).

'Heretofore when a new function was invented, it was for some practical end; to-day they are invented expressly to put at fault the reasonings of our fathers, and one never will get from them anything more than that.

'If logic were the sole guide of the teacher, it would be necessary to begin with the most general functions, that is to say with the most bizarre. It is the beginner that would have to be set grappling with this teratologic museum...' (G. B. Halsted's authorised translation, pp. 435–6). Poincaré discusses the problem with respect to the situation in the theory of real functions – but that does not make any difference.

[1] Paraphrased from Denjoy ([1919], p. 21).

(c) *Improving the conjecture by exception-barring methods. Piecemeal exclusions. Strategic withdrawal or playing for safety*

BETA: I suppose, sir, you are going to explain your puzzling remarks. But, with all apologies for my impatience, I must get this off my chest.

TEACHER: Go on.

(ALPHA *re-enters.*)

BETA: I find some aspects of Delta's arguments silly, but I have come to believe that there is a reasonable kernel to them. It now seems to me that no conjecture is generally valid, but only valid in a certain restricted domain that excludes the *exceptions*. I am against dubbing these exceptions 'monsters' or 'pathological cases'. That would amount to the methodological decision not to consider these as interesting *examples* in their own right, worthy of a separate investigation. But I am also against the term '*counterexample*'; it rightly admits them as examples on a par with the supporting examples, but somehow paints them in war-colours, so that, like Gamma, one panics when facing them, and is tempted to abandon beautiful and ingenious proofs altogether. No: they are just *exceptions*.

SIGMA: I could not agree more. The term 'counterexample' has an aggressive touch and offends those who have invented the proofs. 'Exception' is the right expression. 'There are three sorts of mathematical propositions:

'1. Those which are always true and to which there are neither restrictions nor exceptions, e.g. the angle sum of all plane triangles is always equal to two right angles.

'2. Those which rest on some false principle and so cannot be admitted in any way.

'3. Those which, although they hinge on true principles, nevertheless admit restrictions or exceptions in certain cases...'

EPSILON: What?

SIGMA: '...One should not confuse false theorems with theorems subject to some restriction.'[1] As the proverb says: *The exception proves the rule.*

EPSILON (*to* KAPPA): Who is this muddlehead? He should learn something about logic.

KAPPA (*to* EPSILON): And about non-Euclidean plane triangles.

DELTA: I find it embarrassing to have to predict that in this discus-

---

[1] Bérard [1818–19], p. 347 and p. 349.

sion Alpha and I shall probably be on the same side. We both argued on the basis of a proposition's being either true or false and disagreed only on whether the Euler theorem, in particular, is true or false. But Sigma wants us to admit a third category of propositions that are 'in principle' true but 'admit exceptions in certain cases'. To agree to a peaceful coexistence of theorems and exceptions means to yield to confusion and chaos in mathematics.

ALPHA: *D'accord.*

ETA: I did not want to interfere with the brilliant argumentation of Delta, but now I think it may be profitable if I briefly explain the story of *my* intellectual development. In my schooldays I became – as you would put it – a monster-barrer, not as a defence against Alpha-types but as a defence against Sigma-types. I remember reading in a periodical about the Euler theorem: 'Brilliant mathematicians have put forward proofs of the general validity of the theorem. Nevertheless it suffers exceptions...it is necessary to draw attention to these exceptions since even recent authors do not always recognise them explicitly.'[1] This paper was not an isolated exercise in diplomacy. 'Although in geometry textbooks and lectures it is always pointed out that Euler's beautiful theorem $V + F = E + 2$ is subject to "restriction" in some cases, or "does not seem to be valid", one does not learn the real reason for these exceptions.'[2] Now I looked at the 'exceptions' very carefully and I came to the conclusion that they do not comply with the true definition of the entities in question. So the proof and the theorem can be reinstated and the chaotic coexistence of theorems and exceptions vanishes.

ALPHA: Sigma's chaotic position may serve as an explanation for your monster-barring, but not as an excuse, let alone a justification. Why not eliminate the chaos by accepting the credentials of the counterexample and rejecting the 'theorem' and the 'proof'?

---

[1] Hessel [1832], p. 13. Hessel rediscovered Lhuilier's 'exceptions' in 1832. Just after submitting his manuscript he came across Lhuilier's [1812–13*a*]. He nevertheless decided not to withdraw the paper, most of whose results thus turned out to have already been published, because he thought that the point should be driven home to the 'recent authors' ignoring these exceptions. One of these authors, by the way, happened to be the Editor of the Journal to which Hessel submitted the paper: A. L. Crelle. In his [1826–7] textbook he 'proved' that Euler's theorem was true for *all* polyhedra (vol. 2, pp. 668–71).

[2] Matthiessen ([1863], p. 449). Matthiessen refers here to Heis and Eschweiler's *Lehrbuch der Geometrie* and to Grunert's *Lehrbuch der Stereometrie*. Matthiessen however does not solve the problem – like Eta – by monster-barring, but – like Rho – by monster-adjustment (cf. footnote 2, p. 38).

ETA: Why should I reject the proof? I cannot see anything wrong with it. Can you? My monster-barring seems more rational to me than your proof-barring.

TEACHER: This debate showed that monster-barring may get a more sympathetic audience when it stems from Eta's dilemma. But let us come back to Beta and Sigma. It was Beta who rechristened the counterexamples exceptions. Sigma agreed with Beta...

BETA: I am glad that Sigma agreed with me, but I am afraid that I cannot agree with him. There are certainly three types of propositions: true ones, hopelessly false ones and hopefully false ones. This last type can be improved into true propositions by adding a restrictive clause which states the exceptions. I never 'attribute to formulae an undetermined domain of validity. In reality most of the formulae are true only if certain conditions are fulfilled. By determining these conditions and, of course, pinning down precisely the meaning of the terms I use, I make all uncertainty disappear.'[1] So, as you see, I do not advocate any sort of peaceful coexistence between unimproved formulae and exceptions. I improve my formulae and turn them into *perfect* ones, like those in Sigma's first class. This means that I *accept* the method of monster-barring in so far as it serves for finding *the domain of validity of the original conjecture*; I *reject* it in so far as it functions as a linguistic trick for rescuing 'nice' theorems by restrictive concepts. These two functions of Delta's method should be kept separate. I should like to baptise *my* method, which is characterised by the first of these functions only, '*the exception-barring method*'. I shall use it to determine precisely the domain in which the Euler conjecture holds.

TEACHER: What is the 'precisely determined domain' of Eulerian polyhedra you promised? What is your 'perfect formula'?

BETA: *For all polyhedra that have no cavities (like the pair of nested cubes) and tunnels (like the picture-frame), $V - E + F = 2$.*

TEACHER: Are you sure?

BETA: Yes, I am sure.

TEACHER: What about the twintetrahedra?

BETA: I am sorry. *For all polyhedra that have no cavities, tunnels or 'multiple structure', $V - E + F = 2$.*[2]

---

[1] This is from Cauchy's introduction to his celebrated [1821].

[2] Lhuilier and Gergonne seem to have been sure that Lhuilier's list had enumerated all the exceptions. We read in the introduction to this part of the paper: 'One will easily be convinced that Euler's Theorem is true in general, for all polyhedra, whether they are convex or not, except for those instances that will be specified...' (Lhuilier [1812–13a],

TEACHER: I see. I agree with your policy of improving the conjecture instead of just taking or leaving it. I prefer it both to the method of monster-barring and to that of surrender. However, I have two objections. *First* I contend that your claim that your method not only improves, but 'perfects' the conjecture, that it 'renders it strictly correct', that 'it makes all uncertainties disappear' is untenable.

BETA: Indeed?

TEACHER: You must admit that each new version of your conjecture is only an *ad hoc* elimination of a counterexample which has just cropped up. When you stumble upon nested cubes you exclude polyhedra with *cavities*. When you happen to notice a picture-frame, you exclude polyhedra with *tunnels*. I appreciate your open and observant mind; to take notice of these exceptions is all very well, but I think it would be worth while to inject some method into your blind groping for 'exceptions'. It is good to admit that 'All polyhedra are Eulerian' is only a conjecture. But why give 'All polyhedra without cavities, tunnels and what not are Eulerian' the status of a theorem that is not conjectural any more? How can you be sure that you have enumerated *all* exceptions?

BETA: Can you give one that I did not take into account?

ALPHA: What about my urchin?

GAMMA: And my cylinder?

TEACHER: I do not even need a concrete new 'exception' for my argument. My argument was for the *possibility* of further exceptions.

BETA: You may well be right. One should not just shift one's position whenever a new counterexample turns up. One should not say: 'If no exception occur from phenomena, the conclusion may be pronounced generally. But if at any time afterwards any exception should occur, it may then begin to be pronounced with such exceptions as occur.'[1] Let me think. We first guessed that for *all* polyhedra $V - E + F = 2$, because we found it to be true for cubes, octahedra, pyramids, and prisms. We certainly cannot accept 'this miserable way

---

p. 177). Then we read again in Gergonne's comment: '...the specified exceptions which seem to be the only ones that can occur...' (ibid. p. 188). *But in fact Lhuilier missed the twintetrahedra, which were only noticed twenty years later by Hessel ([1832]).* That some leading mathematicians, even mathematicians with a lively interest in methodology like Gergonne, could believe that one could rely upon the exception-barring method, is noteworthy. The belief is analogous to the 'method of division' in inductive logic, according to which there can be a complete enumeration of possible explanations of a phenomenon, and therefore if we can eliminate all but one by the method of *experimentum crucis*, then this last one is proved.

[1] I. Newton [1717], p. 380.

of inferring from the special to the general'.[1] No wonder exceptions cropped up; it is rather surprising that many more were not found much earlier. To my mind this was because we were mostly occupied with *convex* polyhedra. As soon as other polyhedra entered, our generalisations did not work any more.[2] So instead of barring exceptions piecemeal, I shall draw the borderline modestly, but safely: *All convex polyhedra are Eulerian*.[3] And I hope you will grant that this has nothing conjectural about it: that it is a theorem.

GAMMA: What about my cylinder? It is convex!

BETA: It is a joke!

TEACHER: Let us forget about the cylinder for the moment. We can offer some criticism even without the cylinder. In this new, modified version of the exception-barring method, which Beta devised so briskly in answer to my criticism, piecemeal withdrawal has been replaced by a strategic retreat into a domain hoped to be a stronghold of the conjecture. You are playing for safety. But are you as safe as you claim to be? You still have no guarantee that there will not be any exceptions inside your stronghold. Besides, there is the opposite danger. Could you have withdrawn too radically, leaving lots of Eulerian polyhedra outside the walls? Our original conjecture might

---

[1] Abel [1826a]. His criticism seems to be directed against Eulerian inductivism.

[2] This too is paraphrased from the quoted letter, in which Abel was concerned to eliminate the exceptions to general 'theorems' about functions and thereby establish absolute rigour. The original text (including the previous quotation) is this: 'In Higher Analysis very few propositions are proved with definitive rigour. *One finds everywhere the miserable way of inferring from the special to the general*, and it is a marvel that such procedure leads only rarely to what are called paradoxes. It is really very interesting to look for the reason. In my opinion the reason is to be found in the fact that *analysts have been mostly occupied with functions that can be expressed as power series. As soon as other functions enter – which certainly is rarely the case* – one does not get on any more and as soon as one starts drawing false conclusions, an infinite multitude of mistakes will follow, all supporting each other...' (my italics). Poinsot discovered that inductive generalisations 'often' break down in the theory of polyhedra, just as in number theory: 'Most properties are individual and do not obey any general laws' ([1810], § 45). The intriguing characteristic of this caution towards induction is that it puts down its occasional breakdown to the fact that the universe (of facts, numbers, polyhedra) of course contains miraculous exceptions.

[3] This again is very much in keeping with Abel's method. In the same way Abel restricted the domain of suspect theorems about functions to power-series. In the story of the Euler conjecture this restriction to convex polyhedra was fairly common. Legendre, for instance, after giving his rather general definition of polyhedron (cf. footnote 1, p. 14), presents a proof which on the one hand certainly does not apply to all his general polyhedra, but on the other hand applies to more than convex ones. Nevertheless, in an additional note, in fine print (an afterthought after having stumbled on exceptions never stated?), he withdraws, modestly but safely, to convex polyhedra ([1809], pp. 161, 164, 228).

have been an overstatement, but your 'perfected' thesis looks to me very much like an understatement; yet you still cannot be sure that it is not an overstatement as well.

But I should also like to put forward my *second* objection: your argument forgets about the proof; in guessing the domain of validity of the conjecture, you do not seem to need the proof at all. Surely you do not believe that proofs are redundant?

BETA: I have never said that.

TEACHER: No, you did not. But you discovered that our proof did not prove our original conjecture. Does it prove your improved conjecture? Tell me.

BETA: Well...[1]

ETA: Thank you, sir, for this argument. Beta's embarrassment clearly displays the superiority of the defamed monster-barring method. For we say that the proof proves what it has set out to prove and our answer is unequivocal. We do not allow wayward counterexamples to destroy respectable proofs at liberty, even if they are disguised as meek 'exceptions'.

BETA: I do not find it embarrassing at all that I have to elaborate, improve, and – excuse me, sir – *perfect* my methodology on the stimulus of criticism. My answer is this. I reject the original conjecture as false because there are exceptions to it. I also reject the proof because the same exceptions are exceptions to at least one of the lemmas. (In your terminology this would be: a global counterexample is necessarily also a local counterexample.) Alpha would stop at this point since refutations seem to satisfy his intellectual needs completely. But I go on. By suitably restricting *both* conjecture and proof to the proper domain, I

---

[1] Many working mathematicians are puzzled about what proofs are for if they do not prove. On the one hand they know from experience that proofs are fallible but on the other hand they know from their dogmatist indoctrination that *genuine* proofs must be infallible. *Applied mathematicians* usually solve this dilemma by a shamefaced but firm belief that the proofs of the *pure mathematicians* are 'complete', and so *really* prove. Pure mathematicians, however, know better – they have such respect only for the 'complete proofs' of *logicians*. If asked what is then the use, the function, of their 'incomplete proofs', most of them are at a loss. For instance, G. H. Hardy had a great respect for the logicians' demand for formal proofs, but when he wanted to characterise mathematical proof 'as we working mathematicians are familiar with it', he did it in the following way: 'There is strictly speaking no such thing as mathematical proof; we can, in the last analysis, do nothing but point;...proofs are what Littlewood and I call *gas*, rhetorical flourishes designed to affect psychology, pictures on the board in the lecture, devices to stimulate the imagination of pupils' ([1928], p. 18). R. L. Wilder thinks that a proof is 'only a testing process that we apply to suggestions of our intuition' ([1944], p. 318). G. Pólya points out that proofs, even if incomplete, establish connections between mathematical facts and this helps us to keep them in our memory: proofs yield a mnemotechnic system ([1945], pp. 190–1).

perfect the *conjecture* which will now be *true*, and perfect the basically sound *proof* which will now be *rigorous* and will obviously contain no more false lemmas. For instance we saw that not all polyhedra can be stretched flat onto a plane after having a face removed. But all *convex* polyhedra can. I can rightly call my perfected and rigorously proved conjecture a *theorem*. I state it again: '*All convex polyhedra are Eulerian*.' For convex polyhedra all the lemmas will be manifestly true and the proof, which was not rigorous in its false generality, will be rigorous for the restricted domain of convex polyhedra. So, sir, I have answered your question.

TEACHER: So the lemmas, which once looked manifestly true before the exception was discovered, will again look manifestly true... until the discovery of the next exception. You admit that 'All polyhedra are Eulerian' was guesswork; you admitted just now that 'All polyhedra without cavities and tunnels are Eulerian' was also guesswork; why not admit that 'All convex polyhedra are Eulerian' is guesswork once again!

BETA: Not '*guesswork*' this time, but *insight*!

TEACHER: I abhor your pretentious 'insight'. I respect conscious *guessing*, because it comes from the best human qualities: courage and modesty.

BETA: I proposed a theorem: 'All convex polyhedra are Eulerian.' You offered only a sermon against it. Could you offer a counter-example?

TEACHER: You cannot know that I shall not. You *improved* the original conjecture, but you cannot claim to have *perfected* the conjecture, to have achieved perfect rigour in your proof.

BETA: Can *you*?

TEACHER: I cannot either. But I think that my method of improving conjectures will be an improvement on yours for I shall establish a unity, a real interaction, between proofs and counterexamples.

BETA: I am ready to learn.

(d) The method of monster-adjustment

RHO: Sir, may I get a few words in edgeways?

TEACHER: By all means.

RHO: I agree that we should reject Delta's monster-barring as a general methodological approach, for it doesn't really take 'monsters' seriously. Beta doesn't take his 'exceptions' seriously either, for he merely lists them and then retreats into a safe domain. Thus both these methods are interested only in a limited, privileged field. *My* method does not practise discrimination. I can show that 'on closer examination the exceptions turn out to be only apparent and the Euler theorem retains its validity even for the alleged exceptions'.[1]

TEACHER: Really?

ALPHA: How can my counterexample 3, the 'urchin' (fig. 5), be an ordinary Eulerian polyhedron? It has 12 star-pentagonal faces...

RHO: I don't see any 'star-pentagons'. Don't you see that in actual fact this polyhedron has ordinary *triangular* faces? There are 60 of them. It also has 90 edges and 32 vertices. Its 'Euler characteristic' is 2.[2] The 12 'star-pentagons', their 30 'edges' and 12 'vertices', yielding the 'characteristic' – 6, are only your fancy. Monsters don't exist, only monstrous interpretations. One has to purge one's mind from perverted illusions, one has to learn how to see and how to define correctly what one sees. My method is therapeutic: where you – erroneously – 'see' a counterexample, I teach you how to recognise – correctly – an example. I adjust your monstrous vision...[3]

ALPHA: Sir, please explain *your* method, before Rho brainwashes us.[4]

---

[1] Matthiessen [1863].

[2] The argument that the 'urchin' is 'really' an ordinary, prosaic Eulerian polyhedron with 60 triangular faces, 90 edges and 32 vertices – '*un hexacontaèdre sans épithète*' – was put forward by the staunch champion of the infallibility of the Euler theorem, E. de Jonquières ([1890a], p. 115). The idea of interpreting non-Eulerian star-polyhedra as triangular Eulerian polyhedra does not however stem from Jonquières but has a dramatic history (cf. footnote 4 below).

[3] Nothing is more characteristic of a dogmatist epistemology than its theory of error. For if some truths are manifest, one must explain how anyone can be mistaken about them, in other words, why the truths are not manifest to everybody. According to its particular theory of error, each dogmatist epistemology offers its particular therapeutics to purge minds from error. Cf. Popper [1963a], Introduction.

[4] Poinsot certainly was brainwashed some time between 1809 and 1858. It was Poinsot who rediscovered star-polyhedra, first analysed them from the point of view of Eulerianness and stated that some of them, like our small stellated dodecahedron, do not comply with Euler's formula ([1810]). Now this same Poinsot states categorically in his [1858] that Euler's formula 'is not only true for convex polyhedra, but for any polyhedron whatsoever, including star-polyhedra' (p. 67 – Poinsot uses the term *polyèdres d'espèce supérieure* for star-polyhedra). The contradiction is obvious. What is the explanation? What happened to the star-polyhedral *counterexamples*? The clue is in the first casual-looking sentence of the paper: 'One can reduce the whole theory of polyhedra to the theory of polyhedra with *triangular* faces.' That is, Poinsot–Alpha was brainwashed and turned into Poinsot–Rho: now he sees only triangles where he previously saw star-polygons: now he sees only examples where he previously saw counterexamples. The self-criticism had to be surreptitious, cryptic, because in scientific tradition there are no patterns available for articulating such volte-faces. One also wonders, did he ever come across ring-shaped faces and if so, did he knowingly reinterpret them with his triangular vision?

The change of vision need not always operate in the same direction. For example,

TEACHER: Let him go on.

RHO: I have made my point.

GAMMA: Could you enlarge on your criticism of Delta's method? Both of you exorcised 'monsters'...

RHO: Delta was taken in by your hallucinations. He agreed that your 'urchin' has 12 faces, 30 edges and 12 vertices, and is non-Eulerian. His thesis was that it is not a polyhedron either. But he erred on both counts. Your 'urchin' *is* a polyhedron and *is* Eulerian. But its star-polyhedral interpretation was a *mis*interpretation. If you don't mind, it is not the imprint of the urchin on a healthy, pure mind, but its distorted imprint on a sick mind, twisting in pain.[1]

KAPPA: But how can you distinguish healthy minds from sick ones, rational from monstrous interpretations?[2]

RHO: What puzzles *me* is how you can mix them up!

SIGMA: Do you really think, Rho, that Alpha never noticed that his 'urchin' might be interpreted as a triangular polyhedron? Of course it might. But a closer look reveals that 'these triangles always lie in fives in the same plane and surround a regular pentagon hiding – like their heart – behind a solid angle. Now the five regular triangles together with the inner heart – the regular pentagon – form a so-called "pentagramma" that according to Theophrastus Paracelsus was the sign of health...'[3]

J. C. Becker in his [1869a] – fascinated by the new conceptual framework of simply- and multiply-connected domains (Riemann [1851]) – allowed for ring-shaped polygons but remained blind to star-polygons (p. 66). Five years after this paper – in which he claimed to have brought the problem to a 'definitive' solution – he broadened his vision and recognised star-polygonal and star-polyhedral patterns where he previously saw only triangles and triangular polyhedra ([1874]).

[1] This is part of a Stoic theory of error, attributed to Chrysippos (cf. Aetius [*c.* 150], IV.12.4; also Sextus Empiricus [*c.* 190], I. 249).

According to the Stoics the 'urchin' would be part of external reality, which produces an imprint upon the soul: the *phantasia* or *visum*. A wise man will not give uncritical assent (*synkatathesis* or *adsensus*) to a *phantasia* unless it matures into a clear and distinct idea (*phantasia katalēptikē* or *comprehensio*), which it cannot do if it is false. The system of clear and distinct ideas forms science (*epistēmē*). In our case the imprint of the 'urchin' on Alpha's mind would be the small stellated dodecahedron, while on Rho's mind it would be the triangular hexacontaeder. Rho would claim that Alpha's star-polyhedral vision cannot possibly mature into a clear and distinct idea, obviously since it would upset the 'proved' Euler formula. Thus the star-polyhedral interpretation would fail and the 'only' alternative to it, namely the triangular interpretation, would become clear and distinct.

[2] This is a standard Sceptic criticism of the Stoic claim that they can distinguish *phantasia* from *phantasia katalēptikē* (e.g. Sextus Empiricus [*c.* 190], I. 405).

[3] Kepler [1619], Lib. II. Propositio XXVI.

RHO: Superstition!

SIGMA: And so for the *healthy* mind the secret of the urchin will be revealed: that it is a new, hitherto undreamt-of regular body, with regular faces and equal solid angles, the beautiful symmetry of which might reveal to us the secrets of universal harmony...[1]

ALPHA: Thank you, Sigma, for your defence which again convinces me that opponents are less embarrassing than allies. Of course my polyhedral figure can be interpreted either as a triangular polyhedron or as a star-polyhedron. I am willing to admit both interpretations on a par...

KAPPA: Are you?

DELTA: But surely one of them is the *true* interpretation!

ALPHA: I am willing to admit both interpretations on a par, but one of them will certainly be a global counterexample to Euler's conjecture. Why admit only the interpretation that is 'well-adjusted' to Rho's preconceptions? Anyway, Sir, will you now explain *your* method?

(e) *Improving the conjecture by the method of lemma-incorporation. Proof-generated theorem versus naive conjecture*

TEACHER: Let us return to the picture-frame. I for one recognise it as a genuine global counterexample to the Euler conjecture, as well as a genuine local counterexample to the first lemma of my proof.

GAMMA: Excuse me, Sir – but how does the picture-frame refute the first lemma?

TEACHER: First remove a face and then try to stretch it flat on the blackboard. You will *not* succeed.

ALPHA: To help your imagination, I will tell you that those and only those polyhedra which you can inflate into a sphere have the property that, after a face is removed, you can stretch the remaining part onto a plane.

It is obvious that such a 'spherical' polyhedron is stretchable onto a plane after a face has been cut out; and vice versa it is equally obvious that, if a polyhedron minus a face is stretchable onto a plane, then you can bend it into a round vase which you can then cover with the missing face, thus getting a spherical polyhedron. But our picture-frame can never be inflated into a sphere; but only into a torus.

TEACHER: Good. Now, unlike Delta, I accept this picture-frame as a criticism of the conjecture. I therefore discard the conjecture in its original form as false, but I immediately put forward a modified,

[1] This is a fair exposition of Kepler's view.

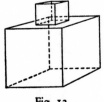

Fig. 12.

restricted version, namely this: the Descartes–Euler conjecture holds good for 'simple' polyhedra, i.e. for those which, after having had a face removed, can be stretched onto a plane. Thus we have rescued some of the original hypothesis. We have: *The Euler characteristic of a simple polyhedron is 2.* This thesis will not be falsified by the nested cube, by the twintetrahedra, or by star-polyhedra – for none of these is 'simple'.

So while the exception-barring method restricted both the domain of the main conjecture and of the guilty lemma to a common domain of safety, thereby accepting the counterexample as criticism both of the main conjecture and of the proof, my method of lemma-incorporation upholds the proof but reduces the domain of the main conjecture to the very domain of the guilty lemma. Or, while a counterexample which is both global and local made the exception-barrer revise both the lemmas and the original conjecture, it makes me revise the original conjecture, but not the lemmas. Do you understand?

ALPHA: Yes, I think I do. To show that I understand, I shall refute you.

TEACHER: My method or my improved conjecture?

ALPHA: Your improved conjecture.

TEACHER: Then you may still not understand my method. But let us have your counterexample.

ALPHA: Consider a cube with a smaller cube sitting on top of it (fig. 12). This complies with all our definitions – *Def. 1, 2, 3, 4, 4', 5* – so it is a genuine polyhedron. And it is 'simple', in that it can be stretched on to the plane. Thus, according to your modified conjecture, its Euler characteristic should be 2. Nonetheless it has 16 vertices, 24 edges and 11 faces, and its Euler characteristic is $16 - 24 + 11 = 3$. It is a gobal counterexample to your improved conjecture and, by the way, also to Beta's first 'exception-barring' theorem. This polyhedron, in spite of having no cavities, tunnels or 'multiple structure', is *not* Eulerian.

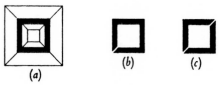

Fig. 13.

DELTA: Let us call this crested cube *Counterexample 6*.[1]

TEACHER: You have falsified my improved conjecture, but you have *not* destroyed my method of improvement. I shall re-examine the proof, and see why it broke down over your polyhedron. There must be another false lemma in the proof.

BETA: Of course there is. I have always suspected the second lemma. It presupposes that in the triangulating process, by drawing a new diagonal edge, you always increase by one the number of edges and of faces. This is false. If we look at the plane network of our crested polyhedron, we shall find a ring-shaped face (fig. 13(*a*)). In this case no single diagonal edge will increase the number of faces (fig. 13(*b*)): we need an increase of two edges to increase the number of faces by one (fig. 13(*c*)).

TEACHER: My congratulations. I certainly must restrict our conjecture further...

BETA: I know what you are going to do. You are going to say that '*Simple polyhedra with triangular faces are Eulerian*'. You will take triangulation for granted; and you will turn this lemma again into a condition.

---

[1] *Counterexample 6* was noticed by Lhuilier ([1812–13*a*], p. 186); Gergonne for once admits the novelty of his discovery! But almost fifty years later Poinsot had not heard of it ([1858]) while Matthiessen ([1863]) and, eighty years later, Jonquières ([1890*b*]) treated it as a monster. (Cf. footnotes 4, p. 31, 2, p. 38.) Primitive exception-barrers of the nineteenth century listed it as a curiosity together with other exceptions: 'As an example one is usually shown the case of a three sided pyramid attached to a face of a tetrahedron so that no edges of the former coincide with an edge of the latter. "Oddly enough, in this case $V - E + F = 3$" is what is written in my college notebook. And that ended the matter' (Matthiessen [1863], p. 449). Modern mathematicians tend to forget about ring-shaped faces, which may be irrelevant for the classification of manifolds but can become relevant in other contexts. H. Steinhaus says in his [1960]: 'Let us divide the globe into F countries (we shall consider *seas* and *oceans* as land). Then we shall have $V + F = E + 2$, whatever the political situation may be' (p. 273). But one wonders whether Steinhaus would destroy West Berlin or San Marino simply because their existence refutes Euler's theorem. (Though of course he may prevent seas like the Baikal from falling completely in one country by defining them as *lakes*, since he has said that only seas and oceans are to be considered as land.)

TEACHER: No, you are mistaken. Before I point out your mistake concretely, let me enlarge upon my comment on your method of exception-barring. When you restrict your conjecture to a 'safe' domain, you do not examine the proof properly, and, in fact, you do not need to for your purpose. The casual statement that in your restricted domain all the lemmas will be true whatever they are, is enough for your purpose. But this is not enough for mine. I build the very same lemma which was refuted by the counterexample *into* the conjecture, so that I have to spot it and formulate it as precisely as possible, on the basis of a careful analysis of the proof. The refuted lemmas thus will be incorporated in my improved conjecture. Your method does not force you to give a painstaking *elaboration of the proof*, since the proof does not appear in your improved conjecture, as it does in mine. Now I return to your present suggestion. The lemma which was falsified by the ring-shaped face was not – as you seem to think – that '*all faces are triangular*' but that '*any face dissected by a diagonal edge falls into two pieces*'. It is *this* lemma which I turn into a condition. Calling the faces which satisfy it 'simply-connected', I can offer a second improvement on my original conjecture: '*For a simple polyhedron, with all its faces simply-connected, $V - E + F = 2$.*' The reason for your rash mis-statement was that your method did not teach you careful proof-analysis. Proof-analysis is sometimes trivial, but sometimes very difficult indeed.

BETA: I see your point. I should also add a self-critical note to your comment, for it seems to me to reveal a whole continuum of exception-barring attitudes. The worst merely bars some exceptions without looking at the proof at all. Hence the mystification when we have the proof on the one hand and the exceptions on the other. In the mind of such primitive exception-barrers, the proof and the exceptions exist in two completely separate compartments. Some others may now point out that the proof will work only in the restricted domain, and thereby claim to dispel the mystery. But their 'conditions' will still be extraneous to the proof-idea.[1] Better exception-

---

[1] '...Lhuilier's memoir consists of two *very distinct* parts. In the first the author offers an original proof of Euler's theorem. In the second his aim is to point out the exceptions to which this theorem is subjected.' (Gergonne's editorial comment on Lhuilier's paper in Lhuilier's [1812–13*a*], p. 172, my italics.)

M. Zacharias in his [1914–31] gives an uncritical but faithful description of this compartmentalisation: 'In the 19th century, geometers, besides finding new proofs of the Euler theorem, were engaged in establishing the exceptions which it suffers under certain conditions. Such exceptions were stated, e.g. by Poinsot. S. Lhuilier and F. Ch. Hessel tried to classify the exceptions...' (p. 1052).

barrers will glance quickly at the proof and gain, as I did just now, some inspiration for stating the conditions which determine a safe domain. The best exception-barrers do a careful analysis of the proof and, on this basis, give a very fine delineation of the prohibited area. In fact your method is, in this respect, a limiting case of the exception-barring method...

IOTA: ...and it displays the fundamental dialectical unity of proof and refutations.

TEACHER: I hope that now all of you see that proofs, even though they may not *prove*, certainly do help to *improve* our conjecture.[1] *The exception-barrers improved it too, but improving was independent of proving. Our method improves by proving. This intrinsic unity between the 'logic of discovery' and the 'logic of justification' is the most important aspect of the method of lemma-incorporation.*

BETA: And of course I now understand your previous puzzling remarks about your not being perturbed by a conjecture being both 'proved' and refuted and about your willingness to 'prove' even a false conjecture.

KAPPA [*aside*]: But why call a 'proof' what in fact is an '*improof*'?

TEACHER: Mind you, few people will share this willingness. Most mathematicians, because of ingrained heuristic dogmas, are incapable of setting out simultaneously to prove *and* refute a conjecture. They would *either* prove it *or* refute it. Moreover, they are particularly incapable of improving conjectures by refuting them if the conjectures happen to be their own. *They want to improve their conjectures without refutations; never by reducing falsehood but by the monotonous increase of truth; thus they purge the growth of knowledge from the horror of counter-examples.* This is perhaps the background to the approach of the best sort of exception-barrers: they *start* by 'playing for safety' by devising a proof for the 'safe' domain and *continue* by submitting it to a thorough critical investigation, testing whether they have made use of each of the imposed conditions. If not, they 'sharpen' or 'generalise' the first modest version of their theorem, i.e. specify the lemmas on which the proof hinges, and incorporate them. For instance, after one or two counterexamples they may formulate the *provisional exception-barring theorem*: 'All convex polyhedra are Eulerian', postponing non-convex instances for a *cura posterior*; next they devise Cauchy's proof and then, discovering that convexity was not really 'used' in the proof, they

[1] Hardy, Littlewood, Wilder and Pólya seem to have missed this point (see footnote 1, p. 29).

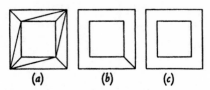

Fig. 14. Three versions of the ring-shaped face: (a) Jonquières,
(b) Matthiessen, (c) the 'untrained eye'.

build up the lemma-incorporating theorem![1] There is nothing
heuristically unsound about this procedure which combines *provisional*
exception-barring with successive proof-analysis and lemma-incorporation.

BETA: Of course this procedure does not abolish criticism, it only
pushes it into the background: instead of directly criticising an overstatement, they criticise an under-statement.

TEACHER: I am delighted, Beta, that I convinced you. Rho and
Delta, how do *you* feel about it?

RHO: I for one certainly think that the problem of 'ring-shaped
faces' is a pseudoproblem. It stems from a monstrous interpretation of
what constitute the faces and edges of this soldering of two cubes into
one – which you called a 'crested cube'.

TEACHER: Explain.

RHO: The 'crested cube' is a polyhedron consisting of two cubes
soldered to one another. Will you agree?

TEACHER: I don't mind.

RHO: Now you misinterpreted 'soldering'. 'Soldering' consists of
edges connecting the vertices of the bottom square of the small cube
to the corresponding vertices of the top square of the large cube. So
there is no 'ring-shaped face' at all.

BETA: The ring-shaped face is there! The dissecting edges you are
talking about are not there!

RHO: They are just hidden from your untrained eyes.[2]

[1] This standard pattern is essentially the one described in the classic of Pólya and Szegö
[1927], p. vii: 'One should scrutinise each proof to see if one has in fact made use of all
the assumptions; one should try to get the same consequence from fewer assumptions
...and one should not be satisfied until counterexamples show that one has arrived at
the boundary of the possibilities.'

[2] This 'soldering' of the two polyhedra by hidden edges is argued by Jonquières ([1890b],
pp. 171–2), who uses monster-barring against cavities and tunnels but monster-adjustment against crested cubes and star-polyhedra. The first proponent of using monster-
adjustment in defence of the Euler theorem was Matthiessen [1863]. He uses monster-

BETA: Do you expect us to take your argument seriously? What *I* see is superstition, but *your* 'hidden' edges are reality?

RHO: Look at this salt crystal. Would you say this is a cube?

BETA: Certainly.

RHO: A cube has 12 edges, hasn't it?

BETA: Yes, it has.

RHO: But on this cube there are no edges at all. They are hidden. They appear only in your rational reconstruction.

BETA: I shall think about this. One thing is clear. The Teacher criticised my conceited view that my method leads to certainty, and also for forgetting about the proof. These criticisms apply just as much to your 'monster-adjustment' as to my 'exception-barring'.

TEACHER: Delta, what about you? How would *you* exorcise the ring-shaped face?

DELTA: I would not. You have converted me to your method. I only wonder why you don't make sure and also incorporate the neglected *third* lemma? I propose a fourth, and, I hope, final formulation: 'All polyhedra are Eulerian, which are (a) simple, (b) have each face simply-connected, and (c) are such that the triangles in the plane triangular network, resulting from stretching and triangulating, can be so numbered that, in removing them in the right order, $V - E + F$ will not alter until we reach the last triangle.'[1] I wonder why you did not propose this at once? If you really took your method seriously, you would

adjustment consistently: he succeeds in displaying hidden edges and faces to explain away everything that is non-Eulerian, including polyhedra with tunnels and cavities. While Jonquières' soldering is a complete triangulation of the ring-shaped face, Matthiessen solders with economy, by drawing only the minimal number of edges that split the face into simply-connected sub-faces (fig. 14).

Matthiessen is remarkably confident about his method of turning revolutionary counterexamples into well-adjusted bourgeois Eulerian examples. He claims that 'any polyhedron can be analysed in such a way that it corroborates Euler's theorem...'. He enumerates the alleged exceptions noted by the superficial observer and then states: 'In each such case we can show that the polyhedron has hidden faces and edges, which, if counted, leave the theorem $V - E + F = 2$ untarnished even for these seemingly recalcitrant cases.'

The idea that, by drawing additional edges or faces, some non-Eulerian polyhedra can be transformed into Eulerian ones, stems however not from Matthiessen, but from Hessel. Hessel illustrates this point with three examples using nice figures ([1832], pp. 14–15). But he did not use this method to 'adjust' but, on the contrary, to 'elucidate the exceptions' by showing 'rather similar polyhedra for which Euler's law is valid'.

[1] This last lemma is unnecessarily strong. It would be enough for the purpose of the proof to replace it by the lemma that 'for the plane triangular network resulting from stretching and triangulating $V - E + F = 1$'. Cauchy does not seem to have noticed the difference.

have turned *all* the lemmas *immediately* into conditions. Why this 'piecemeal engineering'?[1]

ALPHA: Tory turned into revolutionary! Your suggestion strikes me as rather Utopian. For there aren't just *three* lemmas. Why not add, with many others, conditions like '(4) if $1+1 = 2$', and '(5) if all triangles have three vertices and three edges', since we certainly use these lemmas? I propose that we turn only those lemmas into conditions for which a counterexample has been found.

GAMMA: This seems to me too accidental to be accepted as a methodological rule. Let us build in all those lemmas against which we can *expect* counterexamples, i.e. which are not obviously, indubitably true.

DELTA: Well, does our third lemma strike anyone as obvious? Let us turn it into a third condition.

GAMMA: What if the operations expressed by the lemmas of our proof are not all independent? If some of the operations can be performed, it may be that the rest must *necessarily* be able to be performed. I, for one, suspect that *if a polyhedron is simple then there always exists an order of deletion of triangles in the resulting flat network such that $V - E + F$ will not alter*. If there is, then incorporating the first lemma into the conjecture would exempt us from incorporating the third.

DELTA: You claim that the first condition implies the third. Can you prove this?

EPSILON: I can.[2]

ALPHA: The actual proof, however interesting, will not help us in solving our problem: how far should we go in improving our conjecture? I may admit that you have the proof you claim to have – but that will only decompose this third lemma into some new sub-lemmas. Should we now turn these into conditions? Where should we stop?

KAPPA: There is an infinite regress in proofs; therefore proofs do not prove. You should realise that proving is a game, to be played while you enjoy it and stopped when you get tired of it.

EPSILON: No, this is no game but a serious matter. The infinite regress can be halted by trivially true lemmas, which need not be turned into conditions.

---

[1] The students are obviously quite knowledgeable about recent social philosophy. The term was coined by K. R. Popper ([1957], p. 64).

[2] Actually, such a proof was first proposed by H. Reichardt ([1941], p. 23). Also cf. B. L. van der Waerden [1941]. Hilbert and Cohn-Vossen were satisfied that the truth of Gamma's assertion is 'easy to see' ([1932], English translation, p. 292).

GAMMA: This is just what I meant. We do not turn into conditions those lemmas which can be proved from trivially true principles. Nor do we incorporate those lemmas which can be proved – possibly with the help of such trivially true principles – from previously specified lemmas.

ALPHA: Agreed. We can then stop improving our conjecture after we have turned the two non-trivial lemmas into conditions. In fact I do think that this method of improvement, by lemma-incorporation, is flawless. It seems to me that it not only improves but *perfects* the conjecture. And I learned something important from it: that it is wrong to assert that 'the aim of a "problem to prove" is to show conclusively that a certain clearly stated assertion is true, or else to show that it is false'.[1] The *real* aim of a 'problem to prove' should be to *improve* – in fact, perfect – the original, '*naive*' *conjecture* into a genuine '*theorem*'.

Our naive conjecture was 'All polyhedra are Eulerian'.

The monster-barring method defends this naive conjecture by reinterpreting its terms in such a way that at the end we have a *monster-barring theorem*: 'All polyhedra are Eulerian.' But the identity of the linguistic expressions of the naive conjecture and the monster-barring theorem hides, behind surreptitious changes in the meaning of the terms, an essential improvement.

The exception-barring method introduced an element which is really extraneous to the argument: convexity. The *exception-barring theorem* was: 'All convex polyhedra are Eulerian.'

The lemma-incorporating method relied on the argument – i.e. on the proof – and on nothing else. It virtually *summed up the proof in the lemma-incorporating theorem*: 'All simple polyhedra with simply-connected faces are Eulerian.'

This shows that (now I use the term 'proving' in the traditional sense) *one does not prove what one has set out to prove*. Therefore no proof should conclude with the words: '*Quod erat demonstrandum.*'[2]

BETA: Some people say that theorems precede proofs in the order of discovery: 'You have to guess a mathematical theorem before you prove it.' Others deny this, and claim that discovery proceeds by drawing conclusions from a specified set of premises and noting the interesting ones – if you are lucky enough to find any. Or, to use a delightful metaphor of a friend of mine, some say that the heuristic

---

[1] Pólya ([1945], p. 142).
[2] This last sentence is from Alice Ambrose's interesting paper ([1959], p. 438).

'zip fastener' in a deductive structure goes upwards from the bottom – the conclusion – to the top – the premisses,[1] others say that it goes downwards from the top to the bottom. What is your position?

ALPHA: That your metaphor is inapplicable to heuristic. Discovery does not go up or down, but follows a zig-zag path: prodded by counterexamples, it moves from the naive conjecture to the premisses and then turns back again to delete the naive conjecture and replace it by the theorem. Naive conjecture and counterexamples do not appear in the fully fledged deductive structure: the zig-zag of discovery cannot be discerned in the end-product.

TEACHER: Very good. But let us add a note of caution. The theorem does not *always* differ from the naive conjecture. We do not necessarily improve by proving. Proofs improve when the proof-idea discovers unexpected aspects of the naive conjecture which then appear in the theorem. But in *mature* theories this might not be the case. It is certainly the case in young, *growing* theories. This intertwining of discovery and justification, of improving and proving is primarily characteristic of the latter.

KAPPA [*aside*]: Mature theories can be rejuvenated. Discovery always supersedes justification.

SIGMA: This classification corresponds to mine! My first type of propositions was the mature type, the third the growing type...

GAMMA [*interrupts him*]: The theorem is false! I found a counterexample to it.

5. Criticism of the Proof-Analysis by Counterexamples which are Global but not Local. The Problem of Rigour

(a) Monster-barring in defence of the theorem

GAMMA: I have just discovered that my *Counterexample 5*, the cylinder, refutes not only the naive conjecture but also the theorem. Although it satisfies both lemmas, it is not Eulerian.

ALPHA: Dear Gamma, do not become a crank. The cylinder was a joke, not a counterexample. No serious mathematician will take the cylinder for a polyhedron.

GAMMA: Why didn't you protest against my *Counterexample 3*, the

[1] Cf. footnote 1, p. 9. The metaphor of the 'zip fastener' was invented by R. B. Braithwaite; however, he talks only of 'logical' and 'epistemological' zip fasteners, but not of 'heuristic' ones ([1953], esp. p. 352).

urchin? Was that less 'crankish' than my cylinder?[1] *Then* of course you were *criticising* the naive conjecture and welcomed refutations. *Now* you are *defending* the theorem and abhor refutations! *Then,* when a counterexample emerged, your question was: *what is wrong with the conjecture? Now* your question is: *what is wrong with the counterexample?*

DELTA: Alpha, you have turned into a monster-barrer! Aren't you embarrassed?[2]

(b) Hidden lemmas

ALPHA: I am. I may have been a bit rash. Let me think. There are *three possible types of counterexamples.* We have already discussed the *first,* which is local but not global – it certainly would not refute the theorem.[3] The *second,* which is both global and local, does not require any action: far from refuting the theorem, it confirms it. Now there may be a *third* type, which is global but not local. This would refute the theorem. I did not think that this was possible. Now Gamma claims that the cylinder is one. If we do not want to reject it as a monster, we have to admit that it is a global counterexample: $V - E + F = 1$. But is it not of the second harmless type? I bet it does not satisfy at least one of the lemmas.

GAMMA: Let us check. It certainly satisfies the first lemma: if I remove the bottom face, I can easily stretch the rest on to the blackboard.

ALPHA: But if you happen to remove the jacket, the thing falls into two pieces!

GAMMA: So what? The first lemma required that the polyhedron be 'simple', i.e. 'after having had a face removed, it can be stretched on to a plane'. The cylinder satisfies this requirement even if you start by removing the jacket. What you are claiming is that the cylinder should satisfy an *additional* lemma, namely that *the resulting plane network also be connected.* But who has ever stated *this* lemma?

---

[1] The urchin and the cylinder were discussed above, pp. 16 and 31.

[2] Monster-barring in defence of the theorem is an important pattern in informal mathematics: 'What is wrong with the examples in which Euler's formula fails? Which geometrical conditions, rendering more precise the meaning of *F, V,* and *E,* would ensure the validity of Euler's formula?' (Pólya [1954], I, exercise 29). The cylinder is given in exercise 24. The answer is: '...an edge...should terminate in corners...' (p. 225). Pólya formulates this generally: 'The situation, not infrequent in mathematical research is this: A theorem has been already formulated but we have to give a more precise meaning to the terms in which it is formulated in order to render it strictly correct' (p. 55).

[3] Local but not global counterexamples were discussed on pp. 10-12.

ALPHA: Everybody has interpreted 'stretched' as 'stretched in *one piece*', 'stretched *without tear*'...We decided not to incorporate the third lemma because of Epsilon's proof that it followed from the first one.[1] But just have a look at that proof: it hinges on the assumption that the result of the stretching is a *connected* network! Otherwise for the triangulated network $V-E+F$ would not be 1.

GAMMA: Why then didn't you insist on stating it *explicitly*?

ALPHA: Because we took it to be stated *implicitly*.

GAMMA: You, for one, certainly did not. For you proposed that 'simple' stand for 'pumpable into a ball'.[2] The cylinder *can* be pumped into a ball – so according to *your* interpretation it *does* comply with the first lemma.

ALPHA: Well...But you have to agree that it does *not* satisfy the *second* lemma, namely, that '*any face dissected by a diagonal falls into two pieces*'. How will you triangulate the circle or the jacket? Are these faces simply-connected?

GAMMA: Of course they are.

ALPHA: But on the cylinder one cannot draw diagonals at all! A diagonal is an edge that connects two non-adjacent vertices. But your cylinder has no vertices!

GAMMA: Don't get upset. If you want to show that the circle is not simply-connected, draw a diagonal which does *not* create a new face.

ALPHA: Don't be funny; you know very well that I cannot.

GAMMA: Then would you admit that 'there is a diagonal of the circle that does not create a new face' is a *false* statement?

ALPHA: Yes, I would. What are you up to now?

GAMMA: Then you are bound to admit that its negation is true, namely, that 'all diagonals of the circle create a new face', or, that 'the circle is simply-connected'.

ALPHA: You cannot give an *instance* of your lemma that 'all diagonals of the circle create a new face' – therefore it is not *true*, but *meaningless*. Your conception of truth is false.

KAPPA [*aside*]: First they quarrelled about what is a polyhedron, now about what is truth![3]

GAMMA: But you already admitted that the negation of the lemma was *false*! Or can a proposition *A* be *meaningless* while *Not-A* is *meaningful and false*? Your conception of meaning does not make sense!

[1] See p. 40.   [2] See p. 33.

[3] Gamma's *vacuously true statements* were a major innovation of the nineteenth century. Its problem-background has not yet been unfolded.

Mind you, I see your difficulty; but we can overcome it by a slight reformulation. Let us call a face simply-connected if '*for all x, if x is a diagonal then x cuts the face into two*'. Neither the circle nor the jacket can have diagonals, so that in their case, whatever x is, the antecedent will always be false. Therefore the conditional will be instantiated by any object, and will be both meaningful and true. Or, both the circle and the jacket are simply-connected – the cylinder satisfies the second lemma.

ALPHA: No! If you cannot draw diagonals and thereby triangulate the faces, you will never arrive at a flat triangular network and you will never be able to conclude the proof. How can you then claim that the cylinder satisfies the second lemma? Don't you see that *there must be an existential clause* in the lemma? The correct interpretation of the simply-connectedness of a face must be: '*for all x, if x is a diagonal, then x cuts the face into two; and there is at least one x that is a diagonal*'. Our original formulation may not have spelt it out but it was there as an unconsciously made '*hidden assumption*'.[1] All the faces of the cylinder fail to meet it; therefore the cylinder is a counterexample which is *both* global and local, and it does *not* refute the theorem.

GAMMA: First you modified the stretching lemma by introducing 'connectedness', now the triangulating lemma by introducing your existential clause! And all this obscure talk about 'hidden assumptions' only hides the fact that my cylinder made you invent these modifications.

ALPHA: What obscure talk? We already agreed to omit, that is, 'hide', trivially true lemmas.[2] Why then should we state and incorporate trivially *false* lemmas – they are just as trivial and just as boring! Keep them in your mind (*en thyme*) but do not state them. A hidden lemma is not an error: it is shrewd shorthand pointing to our background knowledge.

KAPPA [*aside*]: Background knowledge is where we assume that we know everything but in fact know nothing.[3]

---

[1] 'Euclid...employs an axiom of which he is wholly unconscious' (Russell [1903], p. 407). 'To make [*sic*] a hidden assumption' is a common phrase among mathematicians and scientists. See also Gamow's discussion of Cauchy's proof ([1953], p. 56) or Eves and Newsom on Euclid ([1958], p. 84).

[2] See pp. 40–1.

[3] Good textbooks in informal mathematics usually specify their 'shorthand', i.e. those lemmas, either true or false, which they regard as so trivial as not to be worth mentioning. The standard expression for this is 'we assume *familiarity* with lemmas of type *x*'. The amount of assumed familiarity decreases as criticism turns background knowledge into knowledge. Cauchy, e.g., did not even notice that his celebrated [1821] presupposed

GAMMA: If you did make conscious assumptions, they were that (*a*) removing a face always leaves a connected network and (*b*) any non-triangular face can be dissected into triangles by diagonals. While they were in your *subconscious*, they were listed as *trivially true* – the cylinder however made them somersault into your *conscious* list as *trivially false*. Before being confronted by the cylinder you could not even conceive that the two lemmas could be false. If you now say that you did, then you are rewriting history to purge it from error.[1]

THETA: Not long ago, Alpha, you ridiculed the 'hidden' clauses which cropped up in Delta's definitions after each refutation. Now it is you who make up 'hidden' clauses in the lemmas after each refutation, it is you who shift your ground and try to hide it to save face. Aren't you embarrassed?

KAPPA: Nothing amuses me more than the dogmatist at bay. After donning the militant sceptic's robe to demolish a lesser brand of

'familiarity' with the *theory of real numbers*. He would have rejected as a monster any counterexample which made lemmas about the nature of irrational numbers explicit. Not so Weierstrass and his school: textbooks of informal mathematics now contain a new chapter on the theory of real numbers where these lemmas are collected. But in their introductions 'familiarity with the *theory of rational numbers*' is usually assumed. (See e.g. Hardy's *Pure Mathematics* from the second edition (1914) onwards – the first edition still relegated the theory of real numbers to background knowledge; or Rudin [1953].) More rigorous textbooks narrow down background knowledge even further: Landau, in the introduction to his famous [1930], assumes familiarity only with '*logical reasoning and German language*'. It is ironical that at the very same time Tarski showed that the absolutely trivial lemmas thus omitted may not only be false but inconsistent – German being a semantically closed language. One wonders when 'the author confesses ignorance about the field *x*' will replace the authoritarian euphemism 'the author assumes familiarity with the field *x*': surely only when it is recognised that knowledge has no foundations.

[1] When it is first discovered, the hidden lemma is considered an error. When J. C. Becker first pointed out a 'hidden' (*stillschweigend*) assumption in Cauchy's proof (he quoted the proof second-hand from Baltzer's [1862]), he called it an '*error*' ([1869*a*], pp. 67–8). He drew attention to the fact that Cauchy thought that *all* polyhedra were simple: his lemma was not only hidden but also false. Historians however cannot imagine that great mathematicians should make such errors. A veritable programme of how to falsify history can be found in Poincaré's [1908]: 'A demonstration which is not rigorous is nothingness. I think no one will contest this truth. But if it were taken too literally, we should be led to conclude that before 1820, for example, there was no mathematics: this would be manifestly excessive; the geometers of that time understood voluntarily what we explain by prolix discourse. This does not mean that they did not see it at all; but they passed over it too rapidly, and to see it well would have necessitated taking the pains to say it' (p. 374). Becker's report about Cauchy's 'error' had to be rewritten 1984-wise: 'doubleplusungood refs unerrors rewrite fullwise'. The rewriting was done by E. Steinitz who insisted that 'the fact that the theorem was not generally valid could not possibly remain unnoticed' ([1914–31], p. 20). Poincaré himself applied his programme to the Euler-theorem: 'It is known that Euler proved that $V - E + F = 2$ for *convex* polyhedra' ([1893]) – Euler of course stated his theorem for *all* polyhedra.

dogmatism, Alpha becomes frantic when *he* in turn is cornered by the same sort of sceptical arguments. He now plays fast and loose: trying to fight off Gamma's counterexample first with the defence-mechanism he himself had exposed and forbidden (monster-barring), then by smuggling a reserve of 'hidden lemmas' into the proof and corresponding 'hidden conditions' into the theorem. What is the difference?

TEACHER: The trouble with Alpha was certainly the dogmatist turn in his interpretation of lemma-incorporation. He thought that a careful inspection of the proof would yield a perfect proof-analysis containing *all* the false lemmas (just as Beta thought he could enumerate *all* the exceptions). He thought that by incorporating them he could attain not only an improved theorem, but a *perfected* theorem,[1] *without bothering about counterexamples.* The cylinder showed him to be wrong but, instead of admitting it, he now wants to call a proof-analysis complete if it contains all the *relevant* false lemmas.

(c) The method of proof and refutations

GAMMA: I propose to accept the cylinder as a genuine counterexample to the theorem. I invent a new lemma (or lemmas) that will be refuted by it and add the lemma(s) to the original list. This of course is exactly what Alpha did. But instead of 'hiding' them so that they *become* hidden, I announce them publicly.

Now the cylinder which was a puzzling, dangerous global but not local counterexample (the third type) in respect of the old proof-analysis and of the corresponding old theorem, will be a harmless, global *and* local counterexample (the second type) in respect of the new proof-analysis and the corresponding new theorem.

Alpha thought that his classification of counterexamples was absolute – but in fact it was relative to his proof-analysis. As proof-analysis grows, counterexamples of the third type turn into counterexamples of the second type.

LAMBDA: That is right. A proof-analysis is 'rigorous' or 'valid' and the corresponding mathematical theorem true if, and only if, there is no 'third-type' counterexample to it. I call this criterion the *Principle of Retransmission of Falsity* because it demands that global counterexamples be also local: falsehood should be retransmitted from the naive conjecture to the lemmas, from the consequent of the theorem to its antecedent. If a global but not local counterexample violates this principle, we restore it by adding a suitable lemma to the proof-

[1] See p. 30.

analysis. The Principle of Retransmission of Falsity is therefore a *regulative principle* for proof-analysis *in statu nascendi*, and a global but not local counterexample is a fermenting agent in the growth of proof-analysis.

GAMMA: Remember, even before finding a single refutation we managed to pick out three suspicious lemmas and go ahead with the proof-analysis!

LAMBDA: That is true. Proof-analysis may start not only under the pressure of global counterexamples but also when people have already learned to be on guard against 'convincing' proofs.[1]

In the *first case* all global counterexamples appear as counterexamples of the third type, and all the lemmas start their careers as 'hidden lemmas'. They lead us to a gradual build-up of the proof-analysis and so turn one by one into counterexamples of the second type.

In the *second case* – when we are already in a suspicious mood and look out for refutations – we may arrive at an advanced proof-analysis without any counterexamples. Then there are two possibilities. The *first possibility* is that *we succeed* in refuting – by local counterexamples – the lemmas listed in our proof-analysis. We may very well find that these are also global counterexamples.

ALPHA: This is how I discovered the picture-frame: looking for a polyhedron that, after having a face removed, could not be stretched flat onto a plane.

SIGMA: Then not only do refutations act as fermenting agents for proof-analysis, but proof-analysis may act as a fermenting agent for refutations! What an unholy alliance between seeming enemies!

LAMBDA: That is right. If a conjecture seems very plausible or even self-evident, one should prove it: one may find that it hinges on very sophisticated and dubious lemmas. Refuting the lemmas may lead to some unexpected refutation of the original conjecture.

SIGMA: To proof-generated refutations!

GAMMA: Then 'the virtue of a logical proof is not that it compels belief, but that it suggests doubts'.[2]

---

[1] Our class was a rather advanced one – Alpha, Beta, and Gamma suspected three lemmas when no global counterexamples turned up. In actual history proof-analysis came many decades later: for a long period the counterexamples were either hushed up or exorcised as monsters, or listed as exceptions. The heuristic move from the global counterexample to proof-analysis – the application of the Principle of Retransmission of Falsity – was virtually unknown in the informal mathematics of the early nineteenth century.

[2] H. G. Forder [1927], p. viii. Or: 'It is one of the chief merits of proofs that they instil a certain scepticism as to the result proved.' (Russell [1903], p. 360. He also gives an excellent example.)

LAMBDA: But let me come back to the *second possibility*: when we do *not* find any local counterexamples to the suspected lemmas.

SIGMA: That is, when refutations do not assist proof-analysis! What would happen then?

LAMBDA: We would be branded cranks. The proof would acquire absolute respectability and the lemmas would shake off suspicion. Our proof-analysis would soon be forgotten.[1] Without refutations one cannot sustain suspicion: the searchlight of suspicion soon switches off if a counterexample does not reinforce it, directing the limelight of refutation onto a neglected aspect of the proof that had scarcely been noticed in the twilight of 'trivial truth'.

All this shows that one cannot put proof and refutations into separate compartments. This is why I would propose to rechristen our

---

[1] It is well known that *criticism* may cast doubt on, and eventually refute, '*a priori* truths' and so turn *proofs* into mere *explanations*. That *lack of criticism or of refutation* may turn implausible conjectures into '*a priori* truths' and so tentative explanations into proofs is not so well known but just as important. Two major examples of this pattern are the emergence and fall of Euclid and Newton. The story of their fall is well known, but the story of their emergence is usually misrepresented.

Euclid's geometry seems to have been proposed as a cosmological theory (cf. Popper [1952], pp. 187-9). Both its 'postulates' and 'axioms' (or 'common notions') were proposed as bold, provocative propositions, challenging Parmenides and Zeno, whose doctrines entailed not only the falsity, but even the logical falsity, the inconceivability, of these 'postulates'. Only later were the 'postulates' taken to be indubitably true and the bold anti-Parmenidean 'axioms' (such as 'the whole is greater than the part') taken to be so trivial that they were omitted in later proof-analysis and turned into 'hidden lemmas'. This process started with Aristotle: he branded Zeno a quarrelsome crank, and his arguments 'sophistry'. This story was recently unfolded in exciting detail by Árpád Szabó ([1960], pp. 65-84). Szabó showed that in Euclid's time the word 'axiom' – like 'postulate' – meant a proposition in the critical dialogue (dialectic) put forward to be tested for consequences *without* being admitted as true by the discussion-partner. It is the irony of history that its meaning was turned upside down. The peak of Euclid's authority was reached in the Age of Enlightenment. Clairaut urges his colleagues not to 'obscure proofs and disgust readers' by stating evident truths: Euclid did so only in order to convince 'obstinate sophists' ([1741], pp. x and xi).

Again, *Newton's mechanics and theory of gravitation* was put forward as a daring guess, which was ridiculed and called 'occult' by Leibniz and suspected even by Newton himself. But a few decades later – in the absence of refutations – his axioms came to be taken as indubitably true. Suspicions were forgotten, critics branded 'eccentric' if not 'obscurantist'; some of his most doubtful assumptions came to be regarded as so trivial that textbooks never even stated them. The debate – from Kant to Poincaré – was no longer about the truth of Newtonian theory but about the nature of its certainty. (This volte face in the appraisal of Newtonian theory was first pointed out by Karl Popper – see his [1963a], *passim*.)

The analogy between political ideologies and scientific theories is then more far-reaching than is commonly realised: political ideologies which first may be debated (and perhaps accepted only under pressure) may turn into unquestioned background knowledge even in a single generation: the critics are forgotten (and perhaps executed) until a revolution vindicates their objections.

'*method of lemma-incorporation*' the '*method of proof and refutations*'. Let me state its main aspects in three heuristic rules:

Rule 1. If you have a conjecture, set out to prove it and to refute it. Inspect the proof carefully to prepare a list of non-trivial lemmas (proof-analysis); find counterexamples both to the conjecture (global counterexamples) and to the suspect lemmas (local counterexamples).

Rule 2. If you have a global counterexample discard your conjecture, add to your proof-analysis a suitable lemma that will be refuted by the counterexample, and replace the discarded conjecture by an improved one that incorporates that lemma as a condition.[1] *Do not allow a refutation to be dismissed as a monster.*[2] *Try to make all 'hidden lemmas' explicit.*[3]

Rule 3. If you have a local counterexample, check to see whether it is not also a global counterexample. If it is, you can easily apply Rule 2.

(d) *Proof versus proof-analysis. The relativisation of the concepts of theorem and rigour in proof-analysis.*

ALPHA: What did you mean by 'suitable' in your *Rule 2*?

GAMMA: It is completely redundant. *Any* lemma which is refuted by the counterexample in question can be added – for *any* such lemma will restore the validity of the proof-analysis.

LAMBDA: What! So a lemma like 'All polyhedra have at least 17 edges' would take care of the cylinder! And any other random *ad hoc* conjecture would do just as well, so long as it happened to be refuted by the counterexample.

GAMMA: Why not?

LAMBDA: We already criticised monster-barrers and exception-barrers for forgetting about proofs.[4] Now you are doing the same, inventing a real monster: *proof-analysis without proof*! The only difference between you and the monster-barrer is that you would have Delta make his arbitrary definitions explicit and incorporate them into the theorem as lemmas. And there is *no* difference between exception-barring and your proof-analysing. The only safeguard against such *ad hoc* methods is to use *suitable* lemmas, i.e. lemmas in accordance with

---

[1] This rule seems to have been stated for the first time by P. L. Seidel ([1847], p. 383). See below, p. 136.

[2] 'I have the right to put forward any example that satisfies the conditions of your argument and I strongly suspect that what you call bizarre, preposterous examples are in fact embarrassing examples, prejudicial to your theorem' (G. Darboux [1874b]).

[3] 'I am terrified by the hoard of implicit lemmas. It will take a lot of work to get rid of them' (G. Darboux [1883]).   [4] See pp. 29 and 36.

the spirit of the thought-experiment! Or would you drop the beauty of the proofs from mathematics and replace it by a silly formal game?

GAMMA: Better than your 'spirit of the thought-experiment'! I am defending the objectivity of mathematics against your psychologism.

ALPHA: Thank you, Lambda, you restated my case: one does not *invent* a new lemma out of the blue to cope with a global but not local counterexample: rather, one inspects the proof with increased care and *discovers* the lemma there. So I did not, dear Theta, 'make up' hidden lemmas, and I did not, dear Kappa, 'smuggle' them into the proof. The proof contains all of them – but a mature mathematician understands the entire proof from a brief outline. We should not confuse *infallible proof* with *inexact proof-analysis*. There is still the irrefutable master-theorem: '*All polyhedra on which one can perform the thought-experiment, or briefly, all Cauchy-polyhedra, are Eulerian.*' My approximate proof-analysis drew the borderline of the class of Cauchy-polyhedra with a pencil that – I admit – was not particularly sharp. Now eccentric counterexamples teach us to sharpen our pencil. But first: *no pencil is absolutely sharp* (and if we overdo sharpening it will break); secondly, *pencil-sharpening is not creative mathematics.*

GAMMA: I am lost. What *is* your position? First you were a champion of refutations.

ALPHA: Oh, my growing pains! Mature intuition brushes controversy aside.

GAMMA: Your first mature intuition led you to your 'perfect proof-analysis'. You thought that your 'pencil' was absolutely sharp.

ALPHA: I forgot about the difficulties of linguistic communication – especially with pedants and sceptics. But the heart of mathematics is the thought-experiment – the proof. Its linguistic articulation – the proof-analysis – is necessary for communication but irrelevant. I am interested in polyhedra, you in language. Don't you see the poverty of your counterexamples? They are linguistic, not polyhedral.

GAMMA: Then refuting a theorem only betrays our failure to grasp the hidden lemmas in it? So a 'theorem' is meaningless unless we understand its proof?

ALPHA: Since the vagueness of language makes the *rigour of proof-analysis* unattainable, and turns theorem-formation into an unending process, why bother about the theorem? Working mathematicians certainly do not. If yet another petty 'counterexample' is produced they do not admit that their theorem is refuted, but at most that its 'domain of validity' should be suitably narrowed down.

LAMBDA: So you are not interested either in counterexamples, or in proof-analysis, or in lemma-incorporation?

ALPHA: That is right. I reject all your rules. I propose one single rule instead: *Construct rigorous (crystal-clear) proofs.*

LAMBDA: You argue that the *rigour of proof-analysis* is unattainable. Is *the rigour of proof* attainable? Cannot 'crystal-clear' thought-experiments lead to paradoxical or even contradictory results?

ALPHA: Language is vague, but thought can achieve absolute rigour.

LAMBDA: But surely 'at each stage of evolution our fathers also thought they had reached it? If they deceived themselves, do we not likewise cheat ourselves?'[1]

ALPHA: 'Today absolute rigour is attained.'[2]

[*Giggling in the classroom.*[3]]

GAMMA: This theory of 'crystal-clear' proof is sheer psychologism![4]

ALPHA: Better than the logico-linguistic pedantry of your proof-analysis![5]

LAMBDA: Swearwords apart, I too am sceptical about your conception of mathematics as 'an essentially languageless activity of the

---

[1] Poincaré [1905], p. 214.

[2] *Ibid.* p. 216. Changes in the criterion of 'rigour of the proof' engender major revolutions in mathematics. Pythagoreans held that rigorous proofs have to be arithmetical. However, they discovered a rigorous proof that $\sqrt{2}$ was 'irrational'. When this scandal eventually leaked out, the criterion was changed: arithmetical 'intuition' was discredited and geometrical intuition took its place. This meant a major and complicated reorganisation of mathematical knowledge (e.g. the theory of proportions). In the eighteenth century 'misleading' figures brought geometrical proofs into disrepute, and the nineteenth century saw arithmetical intuition re-enthroned with the help of the cumbersome theory of real numbers. Today the main dispute is about what is rigorous and what not in set-theoretical and metamathematical proofs, as shown by the well-known discussions about the admissibility of Zermelo's and Gentzen's thought-experiments.

[3] As was already pointed out, the class is very advanced.

[4] The term 'psychologism' was coined by Husserl ([1900]). For an earlier 'criticism' of psychologism see Frege [1893], pp. xv–xvi. Modern intuitionists (unlike Alpha) openly embrace psychologism: 'A mathematical theorem expresses a purely empirical fact, namely the success of a certain construction...mathematics...is a study of certain functions of the human mind' (Heyting [1956], pp. 8 and 10). How they reconcile psychologism with certainty is their well-kept secret.

[5] That even if we had perfect knowledge we could not perfectly articulate it, was a commonplace for ancient sceptics (see Sextus Empiricus [c. 190], I. 83–8), but was forgotten in the Enlightenment. It was rediscovered by the intuitionists: they accepted Kant's philosophy of mathematics but pointed out that 'between the perfection of mathematics proper and the perfection of mathematical language no clear connection can be seen' (Brouwer [1952], p. 140). 'Expression by spoken or written word – though necessary for communication – is never adequate...The task of science is not to study languages, but to create ideas' (Heyting [1939], pp. 74–5).

mind'.[1] How can an activity be true or false? Only *articulated* thought can try for truth. Proof cannot be enough: we also have to state what the proof proved. The proof is only a stage of the mathematician's work which has to be followed by proof-analysis and refutations and concluded by the rigorous theorem. We have to *combine* the 'rigour of proof' with the 'rigour of proof-analysis'.

ALPHA: Are you still hoping that at the end you will arrive at a perfectly rigorous proof-analysis? If so, tell me why you did not start by formulating your new theorem 'stimulated' by the cylinder? You only indicated it. Its length and clumsiness would have made us laugh in despair. And this only after the *first* of your new counterexamples! You replaced our original theorem by a succession of ever more precise theorems – but only *in theory*. What about the *practice* of this relativisation? Ever more eccentric counterexamples will be countered by ever more trivial lemmas – yielding a 'vicious infinity'[2] of ever longer and clumsier theorems.[3] Criticism was felt to be invigorating while it seemed to lead to truth. But it is certainly frustrating when it destroys any truth whatsoever and drives us endlessly without purpose. I stop this vicious infinity in *thought* – you will never stop it in *language*.

GAMMA: But I never said that there have to be *infinitely many* counterexamples. At a certain point we may reach truth and then the flow of refutations will stop. But of course we shall not know when. Only refutations are conclusive – proofs are a matter of psychology.[4]

LAMBDA: I still trust that the light of absolute certainty will flash up when refutations peter out!

---

[1] Brouwer [1952], p. 141.

[2] English has the term '*infinite regress*', but this is only a *special* case of 'vicious infinity' (*schlechte Unendlichkeit*) and would not apply here. Alpha obviously coined this phrase with '*vicious circle*' in mind.

[3] Usually mathematicians avoid long theorems by the alternative device of long definitions, so that in the theorems only the defined terms (e.g. 'ordinary polyhedron') appear – this is more economical since one definition abbreviates many theorems. Even so, the definitions take up enormous space in 'rigorous' expositions, though the monsters which lead to them are seldom mentioned. The definition of an '*Euler polyhedron*' (with the definitions of some of the defining terms) takes about 25 lines in Forder [1927] (pp. 67 and 29); the definition of '*ordinary polyhedron*' in the 1962 edition of the *Encyclopaedia Britannica* fills 45 lines.

[4] 'Logic makes us reject certain arguments, but it cannot make us believe any argument' (Lebesgue [1928], p. 328). * *Editors' note:* It should be pointed out that Lebesgue's statement, taken literally, is false. Modern logic has provided us with a precise characterisation of validity, which, it can be shown, some arguments do satisfy. Thus logic certainly can make us believe in an *argument*, though it may not make us believe in the *conclusion* of a valid argument – for we may not believe one or more of the premises.

KAPPA: But will they? What if God created polyhedra so that all true universal statements about them – formulated in human language – are infinitely long? Is it not blasphemous anthropomorphism to assume that (divine) true theorems are of finite length?

Be frank: for some reason or other you are all bored with refutations and piecemeal theorem-formation. Why not call it a day and stop the game? You already gave up 'Quod erat demonstrandum'. Why not give up 'Quod erat demonstratum' too? Truth is only for God.

THETA [aside]: A religious sceptic is the worst enemy of science!

SIGMA: Let's not overdramatise! After all, only a narrow penumbra of vagueness is at stake. It is simply that, as I said before, *not all propositions are true or false.* There is a third class which I would now call '*more or less rigorous*'.

THETA [aside]: Three-valued logic – the end of critical rationality!

SIGMA:...and we state their domain of validity with a rigour that is more or less adequate.

ALPHA: Adequate for what?

SIGMA: Adequate for the solution of the problem which we want to solve.

THETA [aside]: Pragmatism! Has everybody lost interest in *truth*?

KAPPA: Or adequate for the *Zeitgeist*! 'Sufficient unto the day is the rigour thereof.'[1]

THETA: Historicism! [Faints.]

ALPHA: Lambda's rules for '*rigorous proof-analysis*' deprive mathematics of its beauty, present us with the hairsplitting pedantry of long, clumsy theorems filling dull thick books, and will eventually land us in vicious infinity. Kappa's escape-route is convention, Sigma's mathematical pragmatism. What a choice for a rationalist!

GAMMA: So a rationalist ought to relish Alpha's '*rigorous proofs*', inarticulate intuition, 'hidden lemmas', derision of the Principle of Retransmission of Falsity, and elimination of refutations? Should mathematics have no relation to criticism and logic?

BETA: Whatever the case, I am fed up with all this inconclusive verbal quibble. I want to do mathematics and I am not interested in the philosophical difficulties of justifying its foundations. Even if reason fails to provide such justification my natural instinct reassures me.[2]

---

[1] E. H. Moore [1902], p. 411.

[2] 'Nature confutes the sceptics, reason confutes the dogmatists' (Pascal [1659], pp. 1206–7). Few mathematicians would confess – like Beta – that reason is too weak to justify itself. Most of them adopt some brand of dogmatism, historicism or confused pragmatism and remain curiously blind to its untenability; for example: 'Mathematical truths are

I understand Omega has an interesting collection of alternative proofs – I would rather listen to him.

OMEGA: But I shall put them into a 'philosophical' framework!

BETA: I don't mind packing if there is something else in the packet.

Note. In this section I have tried to show how the emergence of mathematical criticism has been the driving force in the search for the 'foundations' of mathematics.

The distinction that we made between *proof* and *proof-analysis* and the corresponding distinction between the *rigour of proof* and the *rigour of proof-analysis* seems to be crucial. About 1800 the *rigour of proof* (crystal-clear thought experiment or construction) was contrasted with muddled argument and inductive generalisation. This was what Euler meant by '*rigida demonstratio*', and Kant's idea of infallible mathematics too was based on this concept (see his paradigm case of a mathematical proof in his [1781], pp. 716–17). It was also thought that one proves what one has set out to prove. It did not occur to anybody that the verbal articulation of a thought-experiment involves any real difficulty. Aristotelian formal logic and mathematics were two completely separate disciplines – mathematicians considered the former as utterly useless. The proof or thought-experiment carried full conviction without any deductive pattern or 'logical' structure.

In the early nineteenth century the flood of counterexamples brought confusion. Since proofs were crystal-clear, refutations had to be miraculous freaks, to be completely segregated from the indubitable proofs. *Cauchy's revolution of rigour* rested on the heuristic innovation that the mathematician should not stop at the proof: he should go on and find out what he has proved by enumerating the exceptions, or rather by stating a safe domain where the proof is valid. *But neither Cauchy – nor Abel – saw any connection between the two problems. It never occurred to them that if they discover an exception, they should have another look at the proof.* (Others practised monster-barring, monster-adjustment or even 'turning a blind eye' – but all agreed that the proof was taboo and had nothing to do with the 'exceptions'.)

The nineteenth-century union of logic and mathematics had two main sources: Non-Euclidean geometry and the *Weierstrassian revolution of rigour.* They brought about the integration of proof (thought-experiment) and refutations and started to develop *proof-analysis*, gradually introducing deductive patterns in the proof-thought-experiment. What we called the 'method of proof and refutations' was their heuristic innovation: *it united logic and mathe-*

in fact *the prototype of the completely incontestable*... But the rigor of maths is not absolute; it is in a process of continual development; the *principles of maths have not congealed once and for all* but have a life of their own and may even be the subject of scientific quarrels.' (A. D. Aleksandrov [1956], p. 7.) (This quotation may remind us that dialectic tries to account for change without using criticism: truths are 'in continual development' but always 'completely incontestable'.)

matics for the first time. Weierstrassian rigour triumphed over its reactionary monster-barring and lemma-hiding opponents who used slogans like 'the dullness of rigour', 'artificiality versus beauty', etc. *The rigour of proof-analysis superseded the rigour of proof:* but most mathematicians put up with its pedantry only so long as it promised them complete certainty.

Cantor's set-theory – with yet another crop of unexpected refutations of 'rigorous' theorems – turned many of the Weierstrassian Old Guard into dogmatists, ever ready to combat the 'anarchists' by barring the new monsters or referring to 'hidden lemmas' in their theorems which represented 'the last word in rigour' while still chastising the older type 'reactionaries' for like sins.

Then some mathematicians realised that the drive for rigour of proof-analysis in the method of proofs and refutations leads to vicious infinity. An 'intuitionist' counter-revolution began: the frustrating logico-linguistic pedantry of *proof-analysis* was condemned, and new extremist standards of rigour were invented for *proofs*; mathematics and logic were divorced once more.

Logicists tried to save the marriage and foundered on the paradoxes. Hilbertian rigour turned mathematics into a cobweb of *proof-analyses* and claimed to stop their infinite regresses by crystal-clear consistency *proofs* of his intuitionistic metatheory. The 'foundational layer', the region of uncriticisable familiarity, was shifted into the thought-experiments of metamathematics. (Cf. Lakatos [1962], pp. 179–84.)

By each 'revolution of rigour' proof-analysis penetrated deeper into the proofs down to the *foundational layer* of 'familiar background knowledge' (also cf. footnote 3, p. 45), where crystal-clear intuition, the rigour of the proof, reigned supreme and criticism was banned. Thus, *different levels of rigour differ only about where they draw the line between the rigour of proof-analysis and the rigour of proof,* i.e. *about where criticism should stop and justification should start.* 'Certainty is never achieved'; 'foundations' are never found – but the 'cunning of reason' turns each increase in *rigour* into an increase in *content,* in the scope of mathematics. But this story is beyond our present investigation.*

---

* *Editors' note.* This historical note, we believe, underplays a little the achievements of the mathematical 'rigorists'. The drive towards 'rigour' in mathematics was, it eventually transpired, a drive towards two separate goals, only one of which is attainable. These two goals are, first, rigorously correct arguments or proofs (in which truth is infallibly transmitted from premises to conclusions) and, secondly, rigorously true axioms, or first principles (which are to provide the original injection of truth into the system – truth would then be transmitted to the whole of mathematics *via* rigorous proofs). The first of these goals turned out to be attainable (given, of course, certain assumptions), whilst the second proved unattainable.

Frege and Russell provided systems into which mathematics could be (fallibly) translated (see below, p. 122), and in which the rules of proof are finite in number and specified in advance. It also turns out that one can show (it is here that the assumptions just referred to come in) that any sentence which can be proved using these rules is a valid consequence of the axioms of the system (i.e. that if these axioms are true, the sentence proved *must* also be true). In these systems there need be no 'gaps' in proofs, and whether a string of sentences is a proof or not can be checked in a finite number of

6. Return to Criticism of the Proof by Counterexamples which are Local but not Global. The Problem of Content

(a) Increasing content by deeper proofs

OMEGA: I like Lambda's method of proof and refutations and I share his faith that somehow we shall finally arrive at a rigorous proof-analysis and thereby at a certainly true theorem. But even so, our very method creates a new problem: *proof-analysis, when increasing certainty, decreases content.* Each new lemma in the proof-analysis, each corresponding new condition in the theorem, reduces its domain. Increasing rigour is applied to a decreasing number of polyhedra. Does lemma-incorporation not repeat the mistake Beta made in playing for safety? Could we too 'have withdrawn too radically, leaving lots of Eulerian polyhedra outside the walls'?[1] In both cases we may throw the baby out with the bathwater. *We should have a counterweight against the content-decreasing pressure of rigour.*

We have already made a few steps in this direction. Let me remind you of two cases and re-examine them.

One was when we first came across local but not global counter-examples.[2] Gamma refuted the third lemma in our first proof-analysis (that 'in removing triangles from the flat triangulated network we have only two possibilities: either we remove an edge or we remove two edges and a vertex'). He removed a triangle from the middle of the network without removing a single edge or vertex.

We then had two possibilities.[3] The *first* was to incorporate the false lemma into the theorem. This would have been a perfectly proper procedure as far as certainty is concerned, but would have reduced the domain of the theorem so drastically that it would have applied only

steps. (Of course, if this checking process shows the sequence of formulae not to be a proof in the system considered, this does not establish that no genuine proof of the end formula exists within the system. Thus, in proof checking, there is an asymmetry which operates in favour of verification and against falsification.) There is no serious sense in which such proofs are fallible. (It is true that it may be that everyone who ever checked some such proof made some inexplicable error, but this is not a serious doubt. It is true that the informal (meta-) theorem that such valid proofs transmit truth may be false – but there is no serious reason to think it is.) But the *axioms* of such systems *are* fallible in a non-trivial sense. The attempt to derive all of mathematics from 'self-evident', 'logical' truths, as is well known, broke down.

[1] Above, p. 28.   [2] For the discussion of this first case see above, pp. 10–12.
[3] Omega seems to ignore a third possibility: Gamma may very well claim that since local but not global counterexamples do not show up any violation of the principle of retransmission of falsity, there is no action to be taken.

for the tetrahedron. Together with the counterexamples we would have thrown out all the examples but one.

This was the rationale behind our adoption of the alternative: instead of *narrowing* the domain of the theorem by lemma-incorporation, we *widened* it by replacing the falsified lemma by an unfalsified one. But this vital pattern for theorem-formation was soon forgotten and Lambda did not bother to formulate it as a heuristic rule. It should be:

Rule 4. If you have a counterexample which is local but not global, try to improve your proof-analysis by replacing the refuted lemma by an unfalsified one.

Counterexamples of the first type (local but not global) may provide an opportunity of *increasing* the content of our theorem which is constantly being *reduced* under the pressure of counterexamples of the third type (global but not local).

GAMMA: *Rule 4* shows up again the weakness of Alpha's now discarded 'perfect proof-analysing intuition'.[1] He would have listed the suspicious lemmas, incorporated them immediately and – without caring for counterexamples – formed near-empty theorems.

TEACHER: Omega, let us hear the second example you promised.

OMEGA: In Beta's proof-analysis the second lemma was that '*all faces are triangular*'.[2] This can be falsified by a number of local but not global counterexamples, e.g. by the cube or the dodecahedron. Therefore you, Sir, replaced it by a lemma which is not falsified by them, namely that '*any face dissected by a diagonal edge falls into two pieces*'. But instead of invoking *Rule 4* you rebuked Beta for 'careless proof-analysis'. You will admit that *Rule 4* is better advice than just 'be more careful'.

BETA: You are right, Gamma, and you also make me understand better 'the method of the best sort of exception-barrers'.[3] They start with a cautious, 'safe' proof-analysis and systematically applying *Rule 4* they gradually build up the theorem without uttering a falsehood. After all, it is a matter of temperament whether one approaches truth through ever false over-statements or through ever true under-statements.

OMEGA: That may be right. But one can interpret *Rule 4* in two ways. Hitherto we considered only the first, weaker interpretation:

[1] Cf. above, p. 47.
[2] For the discussion of this second case cf. above, pp. 35–6.
[3] See above, pp. 37–8.

'one *easily* elaborates, improves the proof by replacing the false lemma by a *slightly modified* one which the counterexample will not refute';[1] all that one needs for this is a 'more careful' inspection of the proof and a 'trifling observation'.[2] On this interpretation *Rule 4* is just local patching *within the framework of the original proof.*

I allow also for the alternative, radical interpretation: to replace the lemma – or possibly all the lemmas – not only by trying to squeeze out the last drop of content from the given proof, but possibly by inventing a completely different, more embracing, *deeper* proof.

TEACHER: For example?

OMEGA: I discussed the Descartes–Euler conjecture earlier with a friend who immediately offered a proof, as follows: let us imagine the polyhedron to be hollow, with a surface made of any rigid material, say cardboard. The edges must be clearly painted on its inside. Let the inside be well illuminated, and let one of the faces be the lens of an ordinary camera – that face from which I can take a snapshot showing all edges and vertices.

SIGMA [*aside*]: A camera in a mathematical proof?

OMEGA: So I get a picture of a plane network, which can be dealt with just like the plane network in your proof. Also in the same way, I can show that, if the faces are simply-connected, $V - E + F = 1$, and adding the lens-face which is invisible on the photo, I get Euler's formula. The main lemma is that there is a face of the polyhedron which, if transformed into the lens of a camera, photographs the inside of the polyhedron so that all the edges and all the vertices are on the film. Now I introduce the following abbreviation: instead of 'a polyhedron which has at least one face from which we can photograph *all* the inside', I shall say 'a quasi-convex polyhedron'.

BETA: So your theorem will be: All quasi-convex polyhedra with simply-connected faces are Eulerian.

OMEGA: For brevity and to give credit to the inventor of this particular proof-idea I would rather say: '*All Gergonne-polyhedra are Eulerian*'.[3]

---

[1] Above, p. 11.   [2] Ibid.

[3] Gergonne's proof is to be found in Lhuilier [1812–13a], pp. 177–9. In the original it could not of course contain photographic devices. It says: 'Take a polyhedron, one of its faces being transparent; and imagine that the eye approaches this face from the outside, so closely, that it can perceive the inside of all the other faces...' Gergonne points out modestly that Cauchy's proof is deeper, it 'has the valuable advantage that it does not assume convexity at all'. (It does not occur to him, however, to ask what it *does* assume.) Jacob Steiner later rediscovered essentially the same proof ([1826]). His attention was then called to Gergonne's priority, so he read Lhuilier's paper with the

GAMMA: But there are many simple polyhedra which, although perfectly Eulerian, are so badly indented that they have no face from which the whole of the inside can be photographed! Gergonne's proof is not deeper than Cauchy's – it is Cauchy's that is deeper than Gergonne's!

OMEGA: Of course! I suppose Teacher knew about Gergonne's proof, found out that it was unsatisfactory by some local but not global counterexample and replaced the optical – photographing – lemma by the wider topological – stretching – lemma. Thereby, he arrived at the *deeper* Cauchy proof, not by a 'careful proof-analysis' followed by a slight alteration, but by a radical, imaginative innovation.

TEACHER: I accept your example – but I did not know about Gergonne's proof. But if you did, why did you not tell us about it?

OMEGA: Because I immediately refuted it by non-Gergonnian polyhedra that are Eulerian.

GAMMA: As I have just said, I too found such polyhedra. But is that a reason for scrapping the proof altogether?

OMEGA: I think so.

TEACHER: Have you heard of Legendre's proof? Would you scrap that too?

OMEGA: I certainly would. It is still less satisfactory: its content is even poorer than Gergonne's proof. His thought-experiment started by mapping the polyhedron with a central projection on to a sphere containing the polyhedron. The radius of the sphere he chose as 1. He chose the centre of the projection so that the sphere will be covered completely, once but only once, by a network of spherical polygons. So his first lemma was that such a point exists. His second lemma was that for the polyhedral network on the sphere $V - E + F = 2$ – but this he succeeded in decomposing into trivially true lemmas of spherical trigonometry. But a point from which such a central projection is possible exists only in convex and a few decent 'almost-convex' polyhedra – a class narrower even than that of 'quasi-convex' polyhedra. But this theorem: '*All Legendre-polyhedra are Eulerian*'[1] differs

list of exceptions but this did not prevent him from concluding his proof with the 'theorem': '*All polyhedra are Eulerian*'. (It was Steiner's paper that provoked Hessel – the Lhuilier of the Germans – to write his [1833].)

[1] *Legendre's proof* can be found in his [1803], but not the proof-generated *theorem*, since proof-analysis and theorem-formation were virtually unknown in the eighteenth century. Legendre first defines polyhedra as solids whose surface consists of polygonal faces (p. 161). Then he proves $V - E + F = 2$ *in general* (p. 228). But there is an exception-barring amendment in a note in fine print on p. 164, saying that only *convex* polyhedra

Fig. 15.

completely from that of Cauchy, but only for the worse. It is 'unfortunately incomplete'.[1] It is a 'vain effort which presupposes conditions on which the Euler theorem does not depend at all. It has to be scrapped and one has to look for more general principles'.[2]

BETA: Omega is right. 'Convexity is to a certain extent accidental for Eulerianness. A convex polyhedron might be transformed, for example by a dent or by pushing in one or more of the vertices, into a non-convex polyhedron with the same configurational numbers. Euler's relation corresponds to something more fundamental than convexity.'[3] And you will never capture that by your 'almost' and 'quasi-' frills.

OMEGA: I thought Teacher had captured it in the topological principles of the Cauchy proof in which *all* the lemmas of Legendre's proof are replaced by completely new ones. But then I stumbled upon a polyhedron that refuted even this proof which is certainly the deepest hitherto.

will be considered. He ignored the almost convex fringe. Poinsot was first, in his [1809], to notice when commenting on Legendre's proof, that the Euler formula 'is valid not only for ordinary convex solids, namely, for those whose surface is cut by a straight line in no more than two points: it also holds for polyhedra with re-entrant angles, provided one can find a point in the interior of the solid which serves as the centre of a sphere on to which one can project the faces of the polyhedron by lines leading from the centre, so that the projected faces do not overlap. This applies to an infinity of polyhedra with re-entrant angles. In fact, Legendre's proof applies, as it stands, to all these additional polyhedra' (p. 46).

[1] E. de Jonquières goes on, again lifting an argument from Poinsot's [1858]: 'In invoking Legendre, and like high authorities, one only fosters a widely spread prejudice that has captured even some of the best intellects: that the domain of validity of the Euler theorem consists only of convex polyhedra' ([1890a], p. 111).

[2] This is from Poinsot ([1858], p. 70).   [3] D. M. Y. Sommerville ([1929], pp. 143-4).

TEACHER: Let us hear about it.

OMEGA: You all remember Gamma's 'urchin' (fig. 7). That was of course non-Eulerian. But not all star-polyhedra are non-Eulerian! Take for instance the 'great stellated dodecahedron' (fig. 15). It consists, like the 'small stellated dodecahedron' of pentagrams, but differently arranged. It has 12 faces, 30 edges and 20 vertices, so that $V - E + F = 2$.[1]

TEACHER: Do you then reject our proof?

OMEGA: I do. The satisfactory proof has to explain the Eulerianness also of the 'great stellated dodecahedron'.

RHO: Why not admit that your 'great stellated dodecahedron' is triangular? Your difficulties are imaginary.

DELTA: I agree. But they are imaginary for a different reason. I have taken to star-polyhedra now: they are fascinating. But they are, I am afraid, essentially different from ordinary polyhedra: therefore one cannot possibly conceive a proof that would explain the Eulerian character of, say, the cube, *and* of the 'great stellated dodecahedron' by one single idea.

OMEGA: Why not? You have no imagination. Would you have insisted after Gergonne's and before Cauchy's proof that concave and convex polyhedra are essentially different: therefore one cannot possibly conceive of a proof that would explain the Eulerian character of convex and concave polyhedra by one single idea? Let me quote from Galileo's *Dialogues*:

SAGREDO: So as you see, all planets and satellites – let us call them all 'planets' – are moving in ellipses.

SALVIATI: I am afraid there are planets moving in parabolas. Look at this stone. I throw it away: it moves along a parabola.

SIMPLICIO: But this stone is not a planet! These are two quite separate phenomena!

SALVIATI: Of course this stone is a planet, only thrown with a less mighty hand than that one which launched the Moon.

SIMPLICIO: Nonsense! How can you dare to pool under one head heavenly and earthly phenomena? One has nothing to do with the other! Of course both may be explained by proofs, but I surely expect the two explanations to be totally different! I cannot imagine a proof which should explain the course of a planet in heaven and a projectile on the earth by one single idea!

SALVIATI: You cannot imagine it but I can devise it...[2]

---

[1] This 'great stellated dodecahedron' had already been devised by Kepler ([1619], p. 53), then independently, by Poinsot ([1810]), who first tested if for Eulerianness. Fig. 15 is copied from Kepler's book.

[2] I was unable to trace this quotation.

TEACHER: Never mind projectiles and planets, Omega, have you succeeded in finding a proof to embrace both ordinary Eulerian polyhedra and Eulerian star-polyhedra?

OMEGA: I have not. But I shall.[1]

LAMBDA: Say you do – what is the matter with Cauchy's proof? You must explain why you reject one proof after the other.

(b) Drive towards final proofs and corresponding sufficient and necessary conditions

OMEGA: You criticised proof-analyses for the breakdown of the *retransmission of falsity* by counterexamples of the *third* type.[2] Now I criticise them for the breakdown of the *transmission* of falsity (or what amounts to the same, the *retransmission of truth*) by counterexamples of the *second* type.[3] A proof must explain the phenomenon of Eulerianness in its entire range.

My quest is not only for *certainty* but also for *finality*. The theorem has to be certain – there must not be any counterexamples *within* its domain; but it has also to be *final*: there must not be any examples *outside* its domain. I want to draw a dividing line between examples and counterexamples, and not just between a safe domain of a few examples on the one hand and a mixed bag of examples and counter-examples on the other.

LAMBDA: Or, you want the conditions of the theorem to be not only sufficient, but also necessary!

KAPPA: Let us imagine then, for the sake of the argument, that you found such a master-theorem: '*All master-polyhedra are Eulerian*'. Do you realise that this theorem will only be 'final' if the converse theorem: '*All Eulerian polyhedra are master-polyhedra*' is certain?

OMEGA: Of course.

KAPPA: That is, if certainty gets lost in vicious infinity, so will finality? You will find at least one Eulerian polyhedron outside the domain of each of your ever deeper proofs.

OMEGA: Of course I know that I cannot solve the problem of finality without solving the problem of certainty. I am sure we shall solve both. We shall stop the infinite spate of counterexamples both of the first and the third types.

TEACHER: Your search for increasing content is very important. But why not accept your second criterion of satisfactoriness – finality –

---

[1] Cf. footnote 1, p. 65.   [2] Global, but not local counterexamples.

[3] Counterexamples which are both global and local.

as a pleasant bonus but not obligatory? Why reject interesting proofs that do not contain both sufficient and necessary conditions? Why regard them as refuted?

OMEGA: Well...[1]

LAMBDA: Whatever the case, Omega certainly convinced me that a single proof may not be enough for the critical improvement of a naive conjecture. Our method should include the radical version of his *Rule 4*, and then it should be called the method of '*proofs and refutations*' instead of '*proof and refutations*'.

MU: Excuse my butting in, I have just translated the results of your discussion into quasi-topological terms: The lemma-incorporating method yielded a contracting sequence of the nested *domains of successive improved theorems*; these domains shrank under the continued attack of global counterexamples in the course of the emergence of hidden lemmas and converged to a *limit*: let us call this limit the '*domain of the proof-analysis*'. If we apply the weaker version of *Rule 4*, this domain can be widened under the continued pressure of local counterexamples. This expanding sequence again will have a limit: I shall call it the '*domain of the proof*'. The discussion then has shown that even this limit domain may be too narrow (perhaps even empty). We may have to devise *deeper* proofs whose domains will form an *expanding* sequence, including more and more recalcitrant Eulerian polyhedra which were local counterexamples to previous proofs. These domains, themselves limit-domains, will converge to the double limit of the '*domain of the naive conjecture*' – which is after all the aim of the inquiry.

The topology of this heuristic space will be a problem for mathematical philosophy: will the sequences be infinite, will they converge at all, attain the limit, may the limit be the empty set?

EPSILON: I have found a deeper proof than Cauchy's which explains also the Eulerianness of Omega's 'great stellated dodecahedron'! [*Passes a note to the Teacher.*]

---

[1] The answer is in the celebrated Pappian heuristic of antiquity which applied only to the discovery of 'final', 'ultimate' truths, i.e. to theorems which contained both necessary and sufficient conditions. For 'problems to prove' the main rule of this heuristic was: 'If you have a conjecture, derive consequences from it. If you arrive at a consequence known to be false, the conjecture was false. If you arrive at a consequence known to be true, reverse the order and, if the conjecture can be thus derived from this true consequence, then it was true.' (Cf. Heath [1925], 1, pp. 138–9.) The principle '*causa aequat effectu*' and the quest for theorems with necessary and sufficient conditions were both in this tradition. It was only in the seventeenth century – when all the efforts to apply Pappian heuristic to modern science had failed – that the quest for certainty came to prevail over the quest for finality.

OMEGA: The final proof! The true essence of Eulerianness will now be revealed!

TEACHER: I am sorry, time is running short: we shall have to discuss Epsilon's very sophisticated proof some other time.[1] All I do see is that it will not be final in Omega's sense. Yes, Beta?

(c) Different proofs yield different theorems

BETA: The most interesting point I have learned from this discussion is that different proofs of the same naive conjecture lead to quite different theorems. *The one Descartes–Euler conjecture is improved by each proof into a different theorem.* Our original proof yielded: '*All Cauchy-polyhedra are Eulerian.*' Now we have learned about two completely different theorems: '*All Gergonne-polyhedra are Eulerian*' and '*All Legendre-polyhedra are Eulerian*'. Three proofs, three theorems with one common ancestor.[2] The usual expression '*different proofs of the Euler theorem*' is then confusing, for it conceals the vital role of proofs in theorem-formation.[3]

[1]* *Editors' note:* The contents of Epsilon's note are revealed below, chapter 2.

[2] There are many other proofs of the Euler conjecture. For a detailed heuristic discussion of Euler's, Jordan's and Poincaré's proofs see Lakatos [1961].

[3] Poinsot, Lhuilier, Cauchy, Steiner, Crelle all thought that the different proofs prove the same theorem: the '*Euler-theorem*'. To quote a characteristic sentence from a standard textbook: 'The theorem stems from Euler, the first proof from Legendre, the second from Cauchy' (Crelle [1827], 2, p. 671).

   Poinsot came very near to noticing the difference when he observed that Legendre's proof applied to more than just ordinary convex polyhedra. (See footnote 1 on p. 60.) But when he then compared Legendre's proof with Euler's proof (that one which was based on cutting off pyramidal corners of the polyhedron and arriving at a final tetrahedron without changing the Euler-characteristic) he gave preference to Legendre's on the ground of 'simplicity' [1858]. 'Simplicity' stands here for the eighteenth-century idea of rigour: clarity in the thought-experiment. It did not occur to him to compare the two proofs for *content*: then Euler's proof would have turned out to be superior. (As a matter of fact, there is nothing wrong with Euler's proof. Legendre applied the subjective standard of contemporary rigour and neglected the objective one of content.)

   Lhuilier – in a surreptitious criticism of this passage (he does not mention Poinsot) – points out that Legendre's simplicity is only 'apparent', for it presumes considerable background knowledge in spherical trigonometry ([1812–13a], p. 171). But Lhuilier too believes that Legendre '*proved the same theorem*' as Euler (ibid. p. 170).

   Jacob Steiner joins him in the appraisal of Legendre's proof and in assuming that all proofs prove the same theorem ([1826]). The only difference is that while according to Steiner all the different proofs prove that '*all polyhedra are Eulerian*', according to Lhuilier all the different proofs prove that '*all polyhedra that have no tunnels, cavities and ringshaped faces are Eulerian*'.

   Cauchy wrote his [1813a] on polyhedra when he was in his early twenties, years before his revolution of rigour, and one cannot take it amiss that he repeats Poinsot's comparison of Euler's and Legendre's proofs in the introduction to the second part of his treatise.

PI: The difference between the different proofs goes much deeper. Only the naive conjecture is about polyhedra. The theorems are about Cauchy-objects, Gergonnian objects, Legendrian objects respectively, but not any more about polyhedra.

BETA: Are you trying to be funny?

PI: No, I shall explain my point. But I would do this in a wider context – I want to discuss *concept-formation* in general.

ZETA: We should rather first discuss *content*. I found Omega's *Rule 4* very weak – even in its radical interpretation.[1]

TEACHER: Right. Let us then first hear Zeta's approach to the problem of content and then wind up our debate with a discussion of concept-formation.

7. The Problem of Content Revisited

(a) The naiveté of the naive conjecture

ZETA: I agree with Omega in deploring the fact that monster-barrers, exception-barrers and lemma-incorporators all strove for certain truth at the expense of content. But his *Rule 4*,[2] demanding deeper proofs of the same naive conjecture, is not enough. Why should our search for content be delimited by the first naive conjecture we stumble upon? Why should the aim of our enquiry be the 'domain of the naive conjecture'?

OMEGA: I don't follow you. Surely our problem was to discover the domain of truth of $V - E + F = 2$?

ZETA: It was not! Our problem was to find out the relation between V, E and F for any polyhedron whatsoever. It was a sheer accident that we first got familiar with polyhedra for which $V - E + F = 2$. But a critical inquiry into these 'Eulerian' polyhedra showed us that there are many more non-Eulerian than Eulerian polyhedra. Why not look for the domain of $V - E + F = -6$, $V - E + F = 28$ or $V - E + F = 0$? Aren't they equally interesting?

He – like most of his contemporaries – did not grasp the difference in depth of different proofs and so could not appreciate the real power of his own proof. He thought he had just given yet *another proof of the very same theorem* – but he was rather eager to stress that he had arrived at a rather trivial generalisation of the Euler-formula to certain aggregates of polyhedra.

Gergonne was the first to appreciate the unrivalled depth of Cauchy's proof (Lhuilier [1812–13*a*], p. 179).

[1] See p. 58.   [2] See above, p. 58.

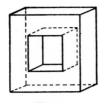

Fig. 16.

SIGMA: You are right. We paid so much attention to $V-E+F = 2$ only because we originally thought it was true. Now we know it is not – we have to find a *new, deeper naive conjecture*...

ZETA:...that will be less naive...

SIGMA:...that will be a relation between V, E, and F for *any* polyhedron.

OMEGA: Why rush? Let us first solve the more modest problem that we set out to solve: to explain why some polyhedra are Eulerian. Until now we have arrived only at partial explanations. For instance, none of the proofs found has explained why a picture-frame with ringshaped faces both in the front and in the back is Eulerian (fig. 16). It has 16 vertices, 24 edges and 10 faces...

THETA: It is certainly not a Cauchy-polyhedron: it has a tunnel, it has ringshaped faces...

BETA: And yet Eulerian! How irrational! Is a polyhedron guilty of a single fault – a tunnel without ringshaped faces (fig. 9) – to be cast out among the goats, yet one which offends in twice as many ways – having also ringshaped faces (fig. 16) – admitted to the sheep?[1]

OMEGA: You see, Zeta, we have enough puzzles about Eulerian polyhedra. Let us solve them before we go on to a more general problem.

ZETA: No, Omega. 'More questions may be easier to answer than just one question. A new more ambitious problem may be easier to handle than the original problem.'[2] Indeed, I shall show you that your narrow, accidental problem can only be solved by solving the wider, essential problem.

OMEGA: But I want to discover the secret of Eulerianness!

ZETA: I understand your resistance. You have fallen in love with

---

[1] The problem was noticed by Lhuilier ([1812–13a], p. 189) and, independently, by Hessel [1832]. In Hessel's paper the figures of the two picture-frames appear next to each other. Also cf. footnote 1, p. 79.

[2] Pólya calls this the 'inventor's paradox' ([1945], p. 110).

the problem of finding out where God drew the boundary dividing Eulerian from non-Eulerian polyhedra. But there is no reason to believe that the term 'Eulerian' occurred in God's blueprint of the universe at all. What if Eulerianness is merely an accidental property of some polyhedra? In this case it would be uninteresting or even impossible to find out the random zig-zags of the demarcation line between Eulerian and non-Eulerian polyhedra. Such an admission however would leave rationalism unsullied, for Eulerianness is then not part of the rational design of the universe. So let us forget about it. One of the main points about critical rationalism is that one is always prepared to abandon one's original problem in the course of the solution and replace it by another one.

(b) Induction as the basis of the method of proofs and refutations

SIGMA: Zeta is right. What a disaster!

ZETA: Disaster?

SIGMA: Yes. You now want a new 'naive conjecture' about the relation between V, E and F, for *any* polyhedron, don't you? Impossible! Look at the vast crowd of counterexamples. Polyhedra with cavities, polyhedra with ringshaped faces, with tunnels, joined together at edges, vertices... $V - E + F$ can take any value whatsoever! You cannot possibly recognise any order in this chaos! We have left the firm ground of Eulerian polyhedra for a swamp! We have irretrievably lost a naive conjecture and have no hope of getting another one!

ZETA: But...

BETA: Why not? Remember the seemingly hopeless chaos in our table of the numbers of vertices, edges and faces even of the most ordinary convex polyhedra.[1] We failed so many times to fit them into a formula.[2] But then suddenly the real regularity governing them struck us: $V - E + F = 2$.

KAPPA [*aside*]: 'Real regularity'? Funny expression for an utter falsehood.

BETA: All that we have to do now is to complete our table with the data for non-Eulerian polyhedra and look for a new formula: with patient, diligent observation, and some luck, we shall hit on the right one; then we can improve it again by applying the method of proofs and refutations!

ZETA: Patient, diligent observation? Trying one formula after the

[1]* *Editors' note:* This table was discussed before we entered the classroom.
[2] See footnote 3, p. 73. The table has been borrowed from Pólya [1954], vol. 1, p. 36.

Polyhedron	F	V	E
I cube	6	8	12
II triangular prism	5	6	9
III pentagonal prism	7	10	15
IV square pyramid	5	5	8
V triangular pyramid	4	4	6
VI pentagonal pyramid	6	6	10
VII octahedron	8	6	12
VIII 'tower'	9	9	16
IX 'truncated cube'	7	10	15

other? Perhaps you will devise a guessing machine that produces random formulas and tests them against your table? Is this your idea of how science progresses?

BETA: I don't understand your scorn. Surely you agree that our first knowledge, our naive conjectures, can only come from diligent observation and sudden insight, however much our critical method of 'proofs and refutations' takes over once we have *found* a naive conjecture? Any deductive method has to start from an inductive basis!

SIGMA: Your inductive method will never succeed. We only arrived at $V - E + F = 2$ because there happened to be no picture-frame or urchin in our original tables. Now that this historical accident...

KAPPA [*aside*]:...or God's benevolent guidance...

SIGMA:...is no more, you will never 'induce' order from chaos. We started with long observation and lucky insight – and failed. Now you propose to start again with longer observation and luckier insight. Even if we did arrive at a new naive conjecture – which I doubt – we shall only end up in the same mess.

BETA: Perhaps we should give up research altogether? We *have* to start again – first with a new naive conjecture and then going again through the method of proofs and refutations.

ZETA: No, Beta. I agree with Sigma – therefore I shall not start again with a new naive conjecture.

BETA: Then where do you want to start if not with an inductive low-level generalisation as a naive conjecture? Or have you an alternative method for starting?

(c) Deductive guessing versus naive guessing

ZETA: Start? Why should I *start*? My mind is not empty when I discover (or invent) a problem.

TEACHER: Do not tease Beta. Here is the problem: '*Is there a relation between the number of vertices, edges and faces of polyhedra analogous to the trivial relation between the number of vertices and edges of polygons, namely that* V = E?'[1] How would *you* set about it?

ZETA: First, I have no government grants to conduct an extensive survey of polyhedra, no army of research assistants counting the numbers of their vertices, edges and faces and compiling tables from the data. But even if I had, I should have no patience – or interest – in trying one formula after the other to test whether it fits.

BETA: What then? Will you lie down on your couch, shut your eyes and forget about the data?

ZETA: Exactly. I need an *idea* to start with, but no data whatsoever.

BETA: And where do you get your idea from?

ZETA: It is already there in our minds when we formulate the problem: in fact, it is in the very formulation of the problem.

BETA: What idea?

ZETA: That for a polygon $V = E$.

BETA: So what?

ZETA: A problem never comes out of the blue. It is always related to our background knowledge. We know that for polygons $V = E$. Now a polygon is a system of polygons consisting of one single polygon. A polyhedron is a system of polygons consisting of more than a single polygon. But for polyhedra $V \neq E$. At what point did the relation $V = E$ break down in the transition from monopolygonal systems to polypolygonal systems? Instead of collecting data I trace how the problem grew out of our background knowledge; or, which was the expectation whose refutation presented the problem?

SIGMA: Right. Let us follow your recommendation. For any polygon $E - V = 0$ (fig. 17(a)). What happens if I fit another polygon to it (not necessarily in the same plane)? The additional polygon has n_1 edges and n_1 vertices; now by fitting it to the original one along a chain of n_1' edges and $n_1' + 1$ vertices we shall increase the number of edges by $n_1 - n_1'$ and the number of vertices by $n_1 - (n_1' + 1)$; that is, in the new 2-polygonal system there will be an excess in the number of edges over the number of vertices: $E - V = 1$ (fig. 17(b)); for an

[1] See above, p. 6.

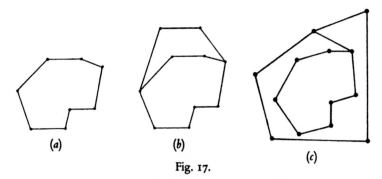

Fig. 17.

unusual but perfectly proper fitting see fig. 17(c)). 'Fitting' a new face to the system will always increase this excess by one, or, for an F-polygonal system constructed in this way $E - V = F - 1$.

ZETA: Or, $V - E + F = 1$.

LAMBDA: But this is false for most polygonal systems. Take a cube...

SIGMA: But my construction can lead only to 'open' polygonal systems – bounded by a circuit of edges! I can easily extend my thought-experiment to 'closed' polygonal systems, with no such boundary. Such closure can be accomplished by covering an open vase-like polygonal system with a polygon-cover: fitting such a covering polygon will increase F by one without changing V or E...

ZETA: Or, for a closed polygonal system – or closed polyhedron – constructed in this way, $V - E + F = 2$: a conjecture which now you have got without 'observing' the number of vertices, edges and faces of a single polyhedron!

LAMBDA: And now you can apply the method of proofs and refutations without an 'inductive starting point'.

ZETA: With the difference that you do not need to devise a proof – the proof is already there! You can go on immediately with refutations, proof-analysis, theorem-formation.

LAMBDA: Then in your method – instead of observations – proof precedes the naive conjecture![1]

ZETA: Well, I shouldn't call a conjecture that has grown out of a proof 'naive'. In my method there is no place for inductive naiveties.

BETA: Objection! You only pushed back the 'naive' inductive start: you start with '$V = E$ for polygons'. Don't you base this on observations?

---

[1] This is an important qualification to footnote 1, p. 9.

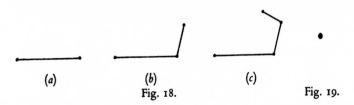

(a)   (b)   (c)

Fig. 18.   Fig. 19.

ZETA: Like most mathematicians, I cannot count. I just tried to count the edges and vertices of a heptagon: I found first 7 edges and 8 vertices, and then again 8 edges and 7 vertices...

BETA: Joking apart, how *did* you get $V = E$?

ZETA: I was deeply shocked when I first realised that for a triangle $V - E = 0$. I knew of course very well that in an edge $V - E = 1$ (fig. 18(a)). I also knew that fitting new edges will always result in an increase by one, both in the number of vertices and edges (figs. 18(b) and 18(c)). Why, in polygonal edge-systems, does $V - E = 0$? Then I realised that this is because of the transition from an open system of edges (which is bounded by two vertices) to a closed system of edges (which has no such boundary): because we 'cover' the open system up by fitting an edge without adding a new vertex. So I proved, not observed, that $V - E = 0$ for polygons.

BETA: Your ingenuity will not help you. You only pushed back the inductive starting point further: now to the statement that $V - E = 1$ for any edge whatsoever. Did you prove or did you *observe* that?

ZETA: I proved it. I knew of course that for a single vertex $V = 1$ (fig. 19). My problem was to construct an analogous relation...

BETA [*furious*]: Didn't you *observe* that for a point $V = 1$?

ZETA: Did *you*? [*Aside, to Pi*]: Should I tell him that my 'inductive starting point' was empty space? That I began by 'observing' *nothing*?

LAMBDA: Whatever the case, two points have been made. First Sigma argued that *it is due only to historical accidents that one can arrive at naive inductive conjectures*: when one is faced with a real chaos of facts, one will scarcely be able to fit them into a nice formula. Then Zeta showed that *for the logic of proofs and refutations we need no naive conjecture, no inductivist starting point at all*.

BETA: Objection! What about those celebrated conjectures that have *not* been preceded (or even followed) by proofs, such as the four-colour conjecture that says that four colours are enough to colour any map, or the Goldbach conjecture? It is only by historical accidents

72

that proofs can precede theorems, that Zeta's 'deductive guessing' can take place: otherwise naive inductive conjectures come first.

TEACHER: We certainly have to learn *both* heuristic patterns: *deductive guessing* is best, but *naive guessing* is better than no guessing at all. But *naive guessing is not induction: there are no such things as inductive conjectures*!

BETA: But we found the naive conjecture by *induction*! 'That is, it was suggested by observation, indicated by particular instances... And among the particular cases that we have examined we could distinguish two groups: those which preceded the formulation of the conjecture and those which came afterwards. The former *suggested* the conjecture, the latter *supported* it. Both kinds of cases provide some sort of contact between the conjecture and "the facts"...'[1] This double contact is the heart of induction: the first makes *inductive heuristic*, the second makes inductive justification, or *inductive logic*.

TEACHER: No! Facts do not suggest conjectures and do not support them either!

BETA: Then what suggested $V - E + F = 2$ to *me*, if not the facts, listed in my table?

TEACHER: I shall tell you. You yourself said you failed many times to fit them into a formula.[2] Now what happened was this: you had three or four conjectures which in turn were quickly refuted. Your table was built up in the process of testing and refuting these conjectures. These dead and now forgotten conjectures suggested the facts, not the facts the conjectures. *Naive conjectures are not inductive conjectures: we arrive at them by trial and error, through conjectures and refutations.*[3] But if you – wrongly – believe that you arrived at them inductively, from your tables, if you believe that the longer the table, the more conjectures it will suggest, and later support, you may waste your time compiling unnecessary data. Also, being indoctrinated that the path of discovery is from facts to conjecture, and from conjecture to proof (the myth of induction), you may completely forget about the heuristic alternative: deductive guessing.[4]

---

[1] Pólya [1954], vol. 1, pp. 5 and 7 (*my italics*).　　　　[2] See pp. 68–9.

[3] These trials and errors are beautifully reconstructed by Pólya. The first conjecture is that F increases with V. This being refuted, two more conjectures follow: E increases with F; E increases with V. The fourth is the winning guess: $F + V$ increases with E ([1954], vol. 1, pp. 35–7).

[4] On the other hand those who, because of the usual deductive presentation of mathematics, come to believe that the path of discovery is from axioms and/or definitions to proofs and theorems, may completely forget about the possibility and importance of naive guessing. In fact in mathematical heuristic it is deductivism which is the greater danger, while in scientific heuristic it is inductivism.

Mathematical heuristic is very like scientific heuristic – not because both are inductive, but because both are characterised by conjectures, proofs, and refutations. The – important – difference lies in the nature of the respective conjectures, proofs (or, in science, explanations), and counterexamples.[1]

BETA: I see. Then our naive conjecture was not the *first* conjecture ever, 'suggested' by hard, non-conjectural facts: it was preceded by many 'pre-naive' conjectures and refutations. The logic of conjectures and refutations has no starting point – but the logic of proofs and refutations has: it starts with the first naive conjecture to be followed by a thought-experiment.

ALPHA: Perhaps. But then I should not have called it 'naive'![2]

KAPPA [*aside*]: Even in heuristic there is no such thing as perfect *naiveté*!

BETA: The main thing is to get out of the trial-and-error period as soon as possible, to proceed quickly to thought-experiments without having too much 'inductive' respect for 'facts'. Such respect may hamper the growth of knowledge. Imagine that you arrive by trial-and-error at the conjecture: $V - E + F = 2$, and that it is immediately refuted by the observation that $V - E + F = 0$ for the picture-frame. If you have too much respect for facts, especially when they refute your conjectures, you will go on with pre-naive trial-and-error and look for another conjecture. But if you have a better heuristic, you at least *try* to ignore the adverse observational test, and try a *test by thought-experiment*: like Cauchy's proof.

SIGMA: What confusion! Why call Cauchy's *proof* a *test*?

BETA: Why call Cauchy's *test* a *proof*? It was a *test*! Listen. You started with a naive conjecture: $V - E + F = 2$ for all polyhedra. Then you drew consequences from it: 'if the naive conjecture is true, after removing a face, for the remaining network $V - E + F = 1$'; 'if this consequence is true, $V - E + F = 1$ even after triangulation'; 'if this last consequence is true, $V - E + F = 1$ will hold while triangles are removed one by one'; 'if this is true, $V - E + F = 1$ for one single triangle'...

---

[1] We owe the revival of mathematical heuristic in this century to Pólya. His stress on the similarities between scientific and mathematical heuristic is one of the main features of his admirable work. What may be considered his only weakness is connected with this strength: he never questioned that science is inductive, and because of his correct vision of deep analogy between scientific and mathematical heuristic he was led to think that mathematics is also inductive. The same thing happened earlier to Poincaré (see his [1902], Introduction) and also to Fréchet (see his [1938]). [2] See above, p. 41.

Now this last conclusion happens to be known to be true. But what if we had concluded that for a single triangle $V - E + F = 0$? We would immediately have rejected the original conjecture as false. All that we have done is to test our conjecture: to draw consequences from it. The test seemed to corroborate the conjecture. But corroboration is not proof.

SIGMA: But then our proof proved even less than we thought it did! We then have to reverse the process and try to construct a thought-experiment which leads in the opposite direction: from the triangle back to the polyhedron!

BETA: That is right. Only Zeta pointed out that instead of solving our problem by first devising a naive conjecture through trial and error, then testing it, then reversing the test into a proof, we can start straight away with the real proof. Had we realised the possibility of deductive guessing we might have avoided all this pseudo-inductive fumbling!

KAPPA [aside]: What a dramatic series of volte-faces! Critical Alpha has turned into a dogmatist, dogmatist Delta into a refutationist, and now inductivist Beta into a deductivist!

SIGMA: But wait. If the *test thought-experiment*...

BETA: I shall call it *analysis*...

SIGMA:...can be followed up at all by a *proof thought-experiment*...

BETA: I shall call it *synthesis*...[1]

SIGMA:...will the 'analytic theorem' be necessarily identical with the 'synthetic theorem'? In going in the opposite direction we might use different lemmas![2]

BETA: If they are different, then the synthetic theorem should supersede the analytic one – after all analysis only *tests* while synthesis *proves*.

TEACHER: Your discovery that our '*proof*' was in fact a *test* seems to have shocked the class and diverted their attention from your main argument: that if we have a conjecture that has already been refuted by a counterexample, we should push the refutation aside and try to test the conjecture by a thought-experiment: this way, we might hit on a proof, leave the phase of trial and error, and switch to the method of proofs and refutations. But it was exactly this which made me say that 'I am willing to set out to "prove" a false conjecture'![3] And

---

[1] According to Pappian heuristic, mathematical discovery starts with a conjecture, which is followed by *analysis* and then, provided *analysis* does not falsify the conjecture, by *synthesis*. (Also cf. above, footnote 1, p. 9, and footnote 1, p. 64.) But while our version of *analysis-synthesis improves* the conjecture, the Pappian version only *proves* or *disproves* it.
[2] Cf. Robinson [1936], p. 471.   [3] See above, p. 23.

Lambda too demanded in his *Rule 1*: 'If you have a conjecture set out to prove it *and* refute it.'

ZETA: That is right. But let me supplement Lambda's rules and Omega's *Rule 4* by

Rule 5. If you have counterexamples of any type, try to find, by deductive guessing, a deeper theorem to which they are counterexamples no longer.

OMEGA: You now stretch my concept of 'depth' – and you may be right. But what about the actual application of your new rule? Until now it has only given us results that we already knew. It is easy to be wise after the event. Your 'deductive guessing' is just the *synthesis* corresponding to Teacher's original *analysis*. But now you should be honest – you must use your method to find a conjecture which you do not already know about, with the promised increase in content.

ZETA: Right. I start with the theorem generated by *my* thought-experiment: '*All closed normal polyhedra are Eulerian.*'

OMEGA: 'Normal'?

ZETA: I don't want to waste time going through the method of proof and refutations. I just call 'normal' all polyhedra that can be built up from a 'perfect' polygon by fitting to it (*a*) first $F-2$ faces without changing $V-E+F$ (these will be *open* normal polyhedra) and (*b*) then a last closing face which increases $V-E+F$ by 1 (and turns the *open* polyhedron into a *closed* one).

OMEGA: 'Perfect polygon'?

ZETA: By a 'perfect' polygon I mean one that can be built up from one single vertex by fitting to it first $n-1$ edges without changing $V-E$, and then a last closing edge which decreases $V-E$ by 1.

OMEGA: Will your closed normal polyhedra coincide with our Cauchy polyhedra?

ZETA: I do not want to go into that now.

(d) Increasing content by deductive guessing

TEACHER: Enough of preliminaries. Let us see your deduction.

ZETA: Yes, Sir. I take two closed normal polyhedra (fig. 20(*a*)) and paste them together along a polygonal circuit so that the two faces that meet disappear (fig. 20(*b*)). Since for the two polyhedra $V-E+F = 4$, the disappearance of two faces in the united polyhedron will just restore the Euler formula – no surprise after Cauchy's proof since the new polyhedron can also easily be pumped into a ball.[1] So the formula

---

[1*] *Editors' note:* This inference is fallacious, although the conclusion is correct. The pasting in fact involves the loss of 8 vertices, 12 edges and 6 faces. The Euler characteristic *is*

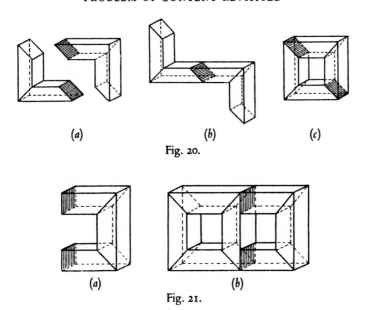

(a)   (b)   (c)

Fig. 20.

(a)   (b)

Fig. 21.

stands up well to this pasting test. But let us now try a double-pasting test: let us 'paste' the two polyhedra together along two polygonal circuits (fig. 20(c)). Now 4 faces will disappear and for the new polyhedron $V-E+F = 0$.

GAMMA: This is Alpha's *Counterexample 4*, the picture-frame!

ZETA: Now if I 'double-paste' to this picture-frame (fig. 20(c)) yet another normal polyhedron (fig. 21(a)), $V-E+F$ will be -2 (fig. 21(b))...

SIGMA: For a monospheroid polyhedron $V-E+F = 2$, for a dispheroid polyhedron $V-E+F = 0$, for a trispheroid $V-E+F = -2$, for an n-spheroid polyhedron $V-E+F = 2-2(n-1)$...

ZETA:...which is your new conjecture of unprecedented content, complete with proof, without having compiled a single table.[1]

SIGMA: This is really nice. Not only did you explain the obstinate picture-frame, but you produced an infinite variety of novel counterexamples...

ZETA: Complete with explanation.

therefore, reduced by two. (The assumed exact coincidence of the two shaded faces in fig. 20 (b) involves reversing the bevelling on one of the half frames so that the broader and the narrower edge are interchanged. Since this operation alters neither V nor E nor F, the argument still, in fact, goes through.)

[1] This was done by Raschig [1891].

RHO: I just arrived at the same result in a different way. Zeta started with two Eulerian examples and turned them into a counterexample in a controlled experiment. I start with a counterexample and turn it into an example. I made the following thought-experiment with a picture-frame: 'Let the polyhedron be of some stuff that is easy to cut like soft clay, let a thread be pulled through the tunnel and then through the clay. It will not fall apart...'[1] But it has become a familiar, simple, spheroid polyhedron! It is true, we increase the number of faces by 2, and the numbers of both edges and vertices by m; but since we know that the Euler characteristic of a simple polyhedron is 2, the original must have had the characteristic 0. Now if one needs more, say n, such cuts to reduce the polyhedron to a simple one, its characteristic will be $2 - 2n$.

SIGMA: This is interesting. Zeta has already shown us that we may not need a conjecture in order to start *proving*, that we may immediately devise a *synthesis*, i.e. a proof thought-experiment from a related proposition that is known to be true. Now Rho shows that we may not need a conjecture even in order to start *testing*, but we may set out – *pretending* that the result is already there – to devise an *analysis*, i.e. a test thought-experiment.[2]

OMEGA: But whichever way you choose, you still leave hordes of polyhedra unexplained! According to your new theorem for all polyhedra $V - E + F$ is an even number, less than 2. But we saw quite a few polyhedra with *odd* Euler characteristics. Take the crested cube (fig. 12) with $V - E + F = 1 \ldots$

ZETA: I never said that my theorem applies to *all* polyhedra. It applies only to all n-spheroid polyhedra built up according to my construction. My construction as it stands does not lead to ringshaped faces.

OMEGA: So?

SIGMA: I know! One can also extend it to polyhedra with ringshaped faces: one may construct a ringshaped polygon by deleting in a suitable proof-generated system of polygons an edge without reducing the number of faces (figs. 22(*a*) and 22(*b*)). I wonder, perhaps there are also 'normal' systems of polygons, constructed in accordance with

[1] Hoppe [1879], p. 102.

[2] This is again part of Pappian heuristic. He calls an *analysis* starting with a conjecture '*theoretical*', and an analysis starting with no conjecture '*problematical*' (Heath [1925], vol. 1, p. 138). The first refers to *problems to prove*, the second to *problems to solve* (or *problems to find*). Also cf. Pólya [1945], pp. 129–36 ('Pappus') and 197–204 ('Working backwards').

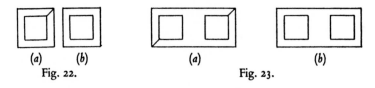

(a)   (b)   (a)   (b)
Fig. 22.   Fig. 23.

our proof, in which we can delete even more than one edge without reducing the number of faces...

GAMMA: That is true. Look at this 'normal' polygonal system (fig. 23(a)). You can delete two edges without reducing the number of faces (fig. 23(b)).

SIGMA: Good! Then in general

$$V - E + F = 2 - 2(n-1) + \sum_{k=1}^{F} e_k$$

for n-spheroid – or n-tuply connected – polyhedra with e_k edges deleted without reduction in the number of faces.

BETA: This formula explains Alpha's crested cube (fig. 12), a monospheroid polyhedron ($n = 1$) with one ringshaped face: e_k are zero, except for e_6 which is 1, or $\sum_{k=1}^{F} e_k = 1$, consequently $V - E + F = 3$.

SIGMA: It also explains your 'irrational' Eulerian freak: the cube with two ringshaped faces and one tunnel (fig. 16). It is a dispheroid polyhedron ($n = 2$) with $\sum_{k=1}^{F} e_k = 2$. Consequently its characteristic is $V - E + F = 2 - 2 + 2 = 2$. Moral order is restored to the world of polyhedra![1]

OMEGA: What about polyhedra with cavities?

---

[1] The 'order' was restored by Lhuilier with approximately the same formula ([1812–13a], p. 189); and by Hessel with clumsy *ad hoc* formulae about different ways of fitting Eulerian polyhedra together ([1832], pp. 19–20). Cf. footnote 1, p. 67.

Historically Lhuilier – in his [1812–13a] – managed to generalise Euler's formula by naive guessing and arrived at the following formula: $V - E + F = 2[(C - T + 1) + (p_1 + p_2 + \ldots)]$, where C is the number of cavities, T the number of tunnels and p_i the number of inner polygons on the ith face. He also *proved* it as far as 'inner polygons' were concerned, but tunnels seem to have defeated him. He constructed the formula in an attempt to account for his three kinds of 'exceptions'; but his list of exceptions was incomplete. (Cf. above, footnote 2, p. 26.) Moreover, this incompleteness was not the only reason for the falsity of his naive conjecture: for he did not notice the possibility that cavities might be multiply-connected; that one may not be able to determine unambiguously the number of tunnels in polyhedra with a system of branching tunnels; and that it is not 'the number of inner polygons', but the number of ringshaped faces,

SIGMA: I know! For them one has to add up the Euler characteristics of each disconnected surface:

$$V - E + F = \sum_{j=1}^{K} \left\{ 2 - 2(n_j - 1) + \sum_{k=1}^{F} e_{kj} \right\}.$$

BETA: And the twintetrahedra?

SIGMA: I know!...

GAMMA: What is the use of all this precision? Stop this flood of pretentious trivialities![1]

ALPHA: Why should he? Or are the twintetrahedra monsters, not genuine polyhedra? A twintetrahedron is just as good a polyhedron as your cylinder! But you liked *linguistic* precision.[2] Why do you deride our new precision? We have to make the theorem cover *all* polyhedra – by making it precise we are increasing its content, not decreasing it. This time precision is a virtue!

KAPPA: Boring virtues are just as bad as boring vices! Besides, you will never achieve *complete* precision. We should stop when it ceases to be interesting to go on.

ALPHA: I have a different point. We started from

(1) one vertex is one vertex.

We deduced from this

(2) $V = E$ for all perfect polygons.

We deduced from this

(3) $V - E + F = 1$ for all normal open polygonal systems.

From this

(4) $V - E + F = 2$ for all normal closed polygonal systems, i.e. polyhedra.

From this again in turn

(5) $V - E + F = 2 - 2(n - 1)$ for normal *n*-spheroid polyhedra.

---

that is relevant (his formula breaks down for two adjacent inner polygons, with an edge in common). For a criticism of Lhuilier's 'inductive generalisation' see Listing [1861], pp. 98–9. Also cf. p. 91, footnote 3.

[1] Quite a few mathematicians of the nineteenth century were confused by such trivial increases in content, and did not really know how to deal with them. Some – like Möbius – used monster-barring definitions (see above, p. 15); others – like Hoppe – monster-adjustment. Hoppe's [1879] is particularly revealing. On the one hand he was keen – like many of his contemporaries – to have a *perfectly complete* 'generalised Euler formula' that covers everything. On the other hand he shrank from trivial complexities. So while he claimed that his formula was 'complete, all-embracing', he added confusedly that 'special cases can make the enumeration (of constituents) dubitable' (p. 103). That is, if an awkward polyhedron still defeats his formula, then its constituents were wrongly counted, and the monster should be adjusted by correct vision: *e.g.* the common vertices and edges of twintetrahedra should be seen and counted twice and each twin recognised as a separate polyhedron (ibid.). For further examples cf. p. 97, footnote 2.   [2] See above, pp. 50–3.

(6) $V - E + F = 2 - 2(n-1) + \sum_{k=1}^{F} e_k$ for normal n-spheroid poly-

hedra with multiply-connected faces.

(7) $V - E + F = \sum_{j=1}^{K} \left\{ 2 - 2(n_j - 1) + \sum_{k=1}^{F} e_{kj} \right\}$ for normal n-spheroid

polyhedra with multiply-connected faces and with cavities.

Isn't this a miraculous unfolding of the hidden riches of the trivial starting-point? And since (1) is indubitably true, so is the rest.

RHO [aside]: Hidden 'riches'? The last two only show how *cheap* generalisations may become![1]

LAMBDA: Do you really think that (1) is the single axiom from which all the rest follows? That deduction increases content?

ALPHA: Of course! Isn't this the miracle of the deductive thought-experiment? If once you have got hold of a little truth, deduction expands it infallibly into a tree of knowledge.[2] If a deduction does *not* increase the content I would not call it deduction, but 'verification': 'verification differs from true demonstration precisely because it is purely analytic and because it is sterile'.[3]

LAMBDA: But surely deduction cannot increase content! If criticism reveals that the conclusion is richer than the premiss, we have to reinforce the premiss by making hidden lemmas explicit.

KAPPA: And it is these hidden lemmas that contain sophistication and fallibility and ultimately destroy the myth of infallible deduction.[4]

TEACHER: Any other question about Zeta's method?

---

[1] Cf. pp. 97-8.

[2] Ancient philosophers did not hesitate to deduce a conjecture from a very trivial consequence of it (see, for example, our synthetic proof leading from the triangle to the polyhedron). Plato thought that 'a single axiom might suffice to generate a whole system'. 'Ordinarily he thought of a single hypothesis as fertile by itself, ignoring in his methodology the other premisses to which he is allying it' (Robinson [1953], p. 168). *This is characteristic of ancient informal logic, that is, of the logic of proof or of thought-experiment or of construction; we regard it as enthymematic only through hindsight: it was only later that an increase in content became a sign, not of the power, but of the weakness, of an inference.* This ancient informal logic was strongly advocated by Descartes, Kant and Poincaré; they all despised Aristotelian formal logic and dismissed it as sterile and irrelevant – at the same time extolling the infallibility of fertile informal logic.

[3] Poincaré [1902], p. 33.

[4] The hunt for hidden lemmas, which started only in mid-nineteenth-century mathematical criticism, was closely related to the process that later replaced *proofs* by *proof-analyses* and *laws of thought* by *laws of language*. The most important developments in logical theory were usually preceded by the development of mathematical criticism. Unfortunately, even the best historians of logic tend to pay exclusive attention to the *changes in logical theory* without noticing their roots in *changes in logical practice*. Cf. also footnote 2, p. 103.

(e) Logical versus heuristic counterexamples.

ALPHA: I like Zeta's *Rule 5*[1] – as I did Omega's *Rule 4*.[2] I liked Omega's method because it looked out for local but not global counterexamples: the ones which Lambda's original three rules[3] ignored as logically harmless, therefore heuristically uninteresting. Omega was stimulated by them to devise new thought-experiments: real advances in our knowledge.

Now Zeta is inspired by counterexamples that are both global and local – perfect corroborations from the logical but not from the heuristic point of view: although corroborations, they still call for action. Zeta proposes to extend, sophisticate our original thought-experiment, to turn logical corroborations into heuristic ones, logically satisfactory instances into instances that are satisfactory from both the logical and the heuristic point of view.

Both Omega and Zeta are for new ideas, while Lambda and especially Gamma are preoccupied with linguistic tricks to deal with their irrelevant global but not local counterexamples – the only relevant ones from their crankish point of view.

THETA: So the logical point of view is 'crankish', is it?

ALPHA: *Your* logical point of view, yes. But I want to make another remark. Whether deduction increases content or not – mind you, of course it does – it certainly seems to guarantee the *continuous growth of knowledge*. We start with a vertex and let knowledge grow forcefully and harmoniously to explain the relation between the number of vertices, edges and faces of any polyhedron whatsoever: an undramatic growth without refutations!

THETA [*to Kappa*]: Has Alpha lost all his judgment? One starts with a *problem*, not with a vertex![4]

ALPHA: This piecemeal but irresistibly victorious campaign will lead us to theorems that are 'not by themselves evident, but only deduced from true and known principles by the continuous and uninterrupted action of a mind that has a clear vision of each step in the process'.[5] They could never have been reached by 'unbiased' observation and a sudden flash of insight.

THETA: I am doubtful about this final victory. Such growth will never bring us to the cylinder – for (1) starts with a vertex and the

---

[1] See p. 76. [2] See p. 58. [3] See p. 50.
[4] Alpha certainly seems to have slipped into the fallacy of deductive heuristic. Cf. p. 73, footnote 4. [5] Descartes [1628], Rule III.

cylinder has none. Also we may never reach one-sided polyhedra, or many-dimensional polyhedra. This piecemeal continuous expansion may well stop at some point and you will have to look for a new, revolutionary start. And even this 'peaceful continuity' is full of refutations, criticism! Why do we go on from (4) to (5), from (5) to (6), from (6) to (7) if not under the continuous pressure of counterexamples which are both global and local? Lambda accepted as genuine counterexamples only those which are global but not local: they revealed the *falsehood* of the theorem. Omega's innovation – rightly praised by Alpha – was to regard also counterexamples which are local but not global as genuine counterexamples: they revealed the *poverty of the truth* of the theorem. Now Zeta tells us to recognise even those counterexamples as genuine which are both global and local: they too point to the *poverty of the truth* of the theorem. For example, picture-frames are both global and local counterexamples to Cauchy's theorem: they are of course corroborations as far as *truth* alone is concerned – but they are refutations as far as *content* is concerned. We may call the first (global but not local) counterexamples *logical*, the others *heuristic counterexamples*. But the more we recognise refutations – logical or heuristic – the quicker knowledge grows. Alpha regards logical counterexamples as irrelevant and refuses to call heuristic counterexamples counterexamples at all, because of his obsession with the idea that growth of mathematical knowledge is continuous, and criticism plays no role.

ALPHA: You expand the concept of refutation and the concept of criticism artificially only to justify your critical theory of the growth of knowledge. Linguistic tricks as tools for a critical philosopher?

PI: I think a discussion of concept-formation may help us to elucidate the issue.

GAMMA: We are all ears.

8. Concept-Formation

(a) *Refutation by concept-stretching. A reappraisal of monster-barring – and of the concepts of error and refutation*

PI: I would first like to go back to the pre-Zeta, or even pre-Omega period, to the three main methods of theorem-formation: monster-barring, exception-barring, and the method of proofs and refutations. Each started with the same naive conjecture, but ended up with different *theorems* and different *theoretical terms*. Alpha has already outlined some aspects of these differences,[1] but his account is unsatisfactory – especially in the case of monster-barring and of the method of proofs and refutations. Alpha thought that the monster-barring theorem 'hides behind the identity of the linguistic expression an essential improvement' on the naive conjecture: he thought that Delta gradually *contracted* the class of 'naive' polyhedra into a class purged of non-Eulerian monsters.

GAMMA: What is wrong with this account?

PI: That it was not the monster-barrers who *contracted* concepts – it was the refutationists who *expanded* them.

DELTA: Hear, hear!

PI: Let us go back to the time of the first explorers of our subject. They were fascinated by the beautiful symmetry of *regular* polyhedra: they thought that the five regular bodies held the secret of the Cosmos. By the time the Descartes–Euler conjecture was put forward, the concept of polyhedron included all sorts of convex polyhedra and even some concave polyhedra. But it certainly did not include polyhedra which were not simple, or polyhedra with ringshaped faces. For the polyhedra that they had in mind, the conjecture *was* true as it stood and the proof was flawless.[2]

Then came the refutationists. In their critical zeal they stretched the concept of polyhedron, to cover objects that were alien to the intended interpretation. The conjecture was true in its *intended interpretation*, it was only false in an *unintended interpretation* smuggled in by the refuta-

---

[1] See p. 41.

[2] Fig. 6 in Euler's [1758a] is the first concave polyhedron ever to appear in a geometrical text. Legendre talks about convex and concave polyhedra in his [1809]. But before Lhuilier nobody mentioned concave polyhedra that were not simple.

However, one interesting qualification might be added. The first class of polyhedra ever investigated consisted partly of the five ordinary regular polyhedra and quasi-regular polyhedra like prisms and pyramids (cf. Euclid). This class was extended after the Renaissance in two directions. One is indicated in the text: to include all convex and some mildly indented simple polyhedra. The other was Kepler's: he widened the class of regular polyhedra by his invention of regular star-polyhedra. But Kepler's innovation was forgotten, only to be made again by Poinsot (cf. above, pp. 16–17). Euler surely did not dream of star-polyhedra. Cauchy knew of them, but his mind was strangely compartmentalised: when he had an interesting idea about star-polyhedra he published it; but he ignored star-polyhedra when presenting counterexamples to his general theorems about polyhedra. Not so the young Poinsot ([1810]) – but later he changed his mind (cf. above, p. 31).

Thus Pi's statement, although heuristically correct (i.e. true in a rational history of mathematics), is historically false. (This should not worry us: actual history is frequently a caricature of its rational reconstructions.)

tionists. Their 'refutation' revealed no *error* in the original conjecture, no *mistake* in the original proof: it revealed the falsehood of a *new* conjecture which nobody had stated or thought of before.

Poor Delta! He valiantly defended the original interpretation of polyhedron. He countered each counterexample with a new clause to safeguard the original concept...

GAMMA: But wasn't it Delta who shifted his position each time? Whenever we produced a new counterexample, he changed his definition for a longer one which displayed another of his 'hidden' clauses!

PI: What a monstrous appraisal of monster-barring! He only *seemed* to shift his position. You wrongly accused him of using surreptitious terminological epicycles in the stubborn defence of an idea. His misfortune was that portentous *Definition 1*: 'A polyhedron is a solid whose surface consists of polygonal faces', which the refutationists seized upon immediately. But Legendre meant it to cover *only* his naive polyhedra; that it covered far more was entirely unrealised and unintended by its proposer. The mathematical public was willing to stomach the monstrous content which slowly emerged from this plausible, innocent-looking definition. This is why Delta had to stutter time and time again, 'I meant...', and had to keep making his endless 'tacit' clauses explicit: all because the naive concept had never been pinned down, and a simple, but monstrous, unintended definition had superseded it. But imagine a different situation, where the definition fixed the intended interpretation of 'polyhedron' properly. Then it would have been up to the refutationists to devise ever longer *monster-including definitions* for, say, 'complex polyhedra': 'A complex polyhedron is an aggregate of (real) polyhedra such that each two of them are soldered by congruent faces.' 'The faces of complex polyhedra can be complex polygons that are aggregates of (real) polygons such that each two of them are soldered by congruent edges.' This *complex polyhedron* would then correspond to Alpha's and Gamma's refutation-generated concept of *polyhedron* – the first definition allowing also for polyhedra that are not simple, the second also for faces that are not simply-connected. So devising new definitions is not necessarily the task of monster-barrers or concept-preservers – it can also be that of monster-includers or concept-stretchers.[1]

SIGMA: Concepts and definitions – that is, intended concepts and

---

[1] An interesting example of monster-including definition is Poinsot's redefinition of *convexity*, which brings star-polyhedra into the respectable class of convex regular bodies [1810].

unintended definitions – can then play funny tricks on each other! I never dreamt that concept-formation might lag behind an unintendedly wide definition!

PI: It might. Monster-barrers only keep to the original concept, while concept-stretchers widen it; the curious thing is that concept-stretching goes on surreptitiously: nobody is aware of it, and since everybody's 'coordinate-system' expands with the widening concept, they fall prey to the heuristic delusion that monster-barring *narrows* concepts, while in fact it keeps them invariant.

DELTA: Now who was intellectually dishonest? Who made surreptitious changes in his position?

GAMMA: I admit we were wrong in indicting Delta for surreptitious contractions of his concept of polyhedron: all his six definitions denoted the same good old concept of polyhedron he inherited from his forefathers. *He defined the very same poor concept in increasingly rich theoretical frames of reference, or languages: monster-barring does not form concepts but only translates definitions*. The monster-barring theorem is no improvement on the naive conjecture.

DELTA: Do you mean that all my definitions were logically equivalent?

GAMMA: That depends on your logical theory – according to mine they certainly are not.

DELTA: This was not a very helpful answer, you will admit. But tell me, did you refute the naive conjecture? You refuted it only by surreptitiously perverting its original interpretation!

GAMMA: Well, we refuted it in a more imaginative and interesting interpretation than you ever dreamt of. This is what makes the difference between *refutations which only reveal a silly mistake* and *refutations which are major events in the growth of knowledge*. If you had found that 'for all polyhedra $V - E + F = 1$' because of inept counting, and I had corrected you, I wouldn't call that a 'refutation'.

BETA: Gamma is right. After Pi's revelation we might hesitate to call our 'counterexamples' *logical counterexamples*, since they are after all not inconsistent with the conjecture in its intended interpretation; but they are certainly *heuristic counterexamples* since they spur the growth of knowledge. If we were to accept Delta's narrow logic, knowledge would not grow. Just suppose that somebody with the narrow conceptual framework discovers the Cauchy proof of the Euler conjecture. He finds that all the steps of this thought-experiment can easily be performed on *any* polyhedron. He takes the 'fact' that all polyhedra are simple and that all faces are simply-connected as obvious, as indubitable. It never occurs to him to turn his 'obvious' lemmas into conditions in an improved conjecture and so to build up a theorem – because the stimulus of counterexamples, in showing up some 'trivially true' lemmas as false, is missing. Thus he thinks that the 'proof' indubitably establishes the truth of the naive conjecture, that its certainty is beyond doubt. But his 'certainty' is far from being a sign of success, it is only a symptom of lack of imagination, of conceptual poverty. It produces smug satisfaction and prevents the growth of knowledge.[1]

[1] This is in fact Cauchy's case. It is likely that had Cauchy already discovered his revolutionary exception-barring method (cf. above, pp. 55–6), he would have searched for and found some exceptions. But he probably came across the problem of exceptions only later, when he decided to clear up the chaos in analysis. (It was Lhuilier who seems to have first noticed, and faced, the fact that such 'chaos' was not confined to analysis.)

Historians, e.g. Steinitz in his [1914–31], usually say that Cauchy, noticing that his theorem was not universally valid, stated it for *convex* polyhedra only. It is true that in his proof he uses the expression 'the convex surface of a polyhedron' ([1813a], p. 81), and in his [1813b] he restates Euler's theorem under the general head: 'Theorems on solid angles and *convex polyhedra*'. But probably to counteract this title, he gives particular stress to the *universal* validity of Euler's theorem for *any* polyhedron (Theorem XI, p. 94), while stating three other theorems (Theorem XIII and its two corollaries) explicitly for *convex* polyhedra (pp. 96 and 98).

Why Cauchy's sloppy terminology? Cauchy's concept of polyhedron *almost* coincided with the concept of convex polyhedron. But it did not coincide exactly: Cauchy knew about concave polyhedra, which can be obtained by slightly pushing in the side of convex polyhedra, but he did not discuss what seemed to be irrelevant further *corroborations* – not *refutations* – of his theorem. (*Corroborations never compare with counterexamples, or even 'exceptions', as catalysts for the growth of concepts.*) This is the reason for Cauchy's casual use of 'convex': it was a failure to realise that concave polyhedra might give counterexamples, not a conscious effort to *eliminate* these counterexamples. In the very same paragraph, he argues that Euler's theorem is an 'immediate consequence' of the lemma that $V - E + F = 1$ for *flat* polygonal networks, and states that 'for the validity of the theorem $V - E + F = 1$ it has no significance whatever whether the polygons lie in the same plane or in different planes, since the theorem is concerned only with the number of polygons and the number of their constituents' (p. 81). This argument is perfectly correct within Cauchy's narrow conceptual framework, but incorrect in a wider one, in which 'polyhedron' refers also to, say, picture-frames. The argument was frequently repeated in the first half of the nineteenth century (e.g. Olivier [1826], p. 230, or Grunert [1827], p. 367, or R. Baltzer [1860–62], vol. 2, p. 207). It was criticised by J. C. Becker ([1869a], p. 68).

Often, as soon as concept-stretching refutes a proposition, the refuted proposition seems such an elementary mistake that one cannot imagine that great mathematicians could have made it. This important characteristic of concept-stretching refutation explains why respectful historians, because they do not understand that concepts *grow*, create for themselves a maze of problems. After saving Cauchy by claiming that he 'could not possibly miss' polyhedra which are not simple and that therefore he '*categorically*' (!) *restricted* the theorem to the domain of convex polyhedra, the respectful historian now has to explain

(b) Proof-generated versus naive concepts. Theoretical versus naive classification

PI: Let me return to the proof-generated theorem: 'All simple *polyhedra* with simply-connected faces are Eulerian'. This formulation is misleading. It should read: 'All simple *objects* with simply-connected faces are Eulerian.'

GAMMA: Why?

PI: The first formulation suggests that the class of simple polyhedra that occurs in the theorem is a subclass of the class of 'polyhedra' of the naive conjecture.

SIGMA: Of course the class of simple polyhedra is a subclass of polyhedra! The concept of 'simple polyhedron' *contracts* the original wide class of polyhedra by restricting them to those on which the first lemma of our proof is performable. The concept of 'simple polyhedron with simply-connected faces' indicates a further contraction of the original class...

PI: No! The original class of polyhedra contained only polyhedra that were simple and whose faces were simply-connected. Omega was wrong when he said that lemma-incorporation reduces content.[1]

OMEGA: But doesn't each incorporation of lemmas rule out a counterexample?

PI: Of course it does: but a counterexample that was produced by concept-stretching.

OMEGA: So lemma-incorporation *conserves* content, just like monster-barring?

PI: No. Lemma-incorporation *increases* content: monster-barring does not.

OMEGA: What? Do you really want to convince me not only that lemma-incorporation does not *reduce* content, but also that it *increases* it? That instead of *contracting* concepts it *stretches* them?

why Cauchy's borderline was 'unnecessarily' narrow. Why did he *ignore* non-convex Eulerian polyhedra? Steinitz's explanation is this: the *correct* formulation of the Euler-formula is in terms of connectivity of surfaces. Since in Cauchy's period this concept was not yet 'clearly grasped', 'the simplest way out' was to assume convexity (p. 20). So Steinitz explains away a mistake that Cauchy never made.

Other historians proceed in a different way. They say that before the point where the correct conceptual framework (i.e. the one they know) was reached there was only a 'dark age' with 'seldom, if ever, sound' results. This point in the theory of polyhedra is Jordan's [1866a] proof according to Lebesgue ([1923], pp. 59–60); it is Poincaré's [1895] according to Bell ([1945], p. 460).

[1] See above, p. 57.

PI: Exactly. Just listen. Was a globe, with a political map drawn on it, an element of the original class of polyhedra?

OMEGA: Certainly not.

PI: But it became one after Cauchy's proof. For you can perform Cauchy's proof on it without the slightest difficulty – if only there are no ringshaped countries or seas on it.[1]

GAMMA: That is right! Pumping the polyhedron up into a ball and distorting edges and faces will not perturb us in the least in performing the proof – so long as the distortion does not alter the *number* of vertices, edges and faces.

SIGMA: I see your point. Then the proof-generated 'simple polyhedron' is not just a contraction, a specification, but also a *generalisation*, an *expansion* of the naive 'polyhedron'.[2] The idea of *generalising* the concept of polyhedron so that it should include crumpled, *curvilinear* 'polyhedra' with *curved* faces could hardly have occurred to anybody before Cauchy's proof; even if it had, it would have been dismissed as crankish. But now it is a natural generalisation, since the operations of our proof can be interpreted for them just as well as for ordinary naive polyhedra with straight edges and flat faces.[3]

PI: Good. But you have to make one more step. *Proof-generated concepts* are neither 'specifications', nor 'generalisations' of naive concepts. The impact of proofs and refutations on naive concepts is

---

[1] Cf. p. 35, footnote 1.

[2] Darboux, in his [1874a], came close to this idea. Later it was clearly formulated by Poincaré: '...mathematics is the art of giving the same name to different things... When the language is well chosen, we are astonished to learn that all the proofs made for a certain object apply immediately to many new objects; there is nothing to change, not even the words, since the names have become the same' ([1908], p. 375). Fréchet calls this 'an extremely useful principle of generalisation', and formulates it as follows: 'When the set of properties of a mathematical entity used in the proof of a proposition about this entity does not determine this entity, the proposition can be extended to apply to a more general entity' ([1928], p. 18). He points out that such generalisations are not trivial and 'may require very great efforts' (ibid.).

[3] Cauchy did not notice this. His proof differed from the one given by the Teacher in one important respect: Cauchy in his [1813a] and [1813b] did *not* imagine the polyhedron to be made of rubber. The novelty of his proof-idea was to imagine the polyhedron as a *surface*, and not as a *solid*, as Euclid, Euler and Legendre did. But he imagined it as a *solid* surface. When he removed one face and mapped the remaining spatial polygonal network into a flat polygonal network, he did not conceive his mapping as a *stretching* that might *bend* faces or edges. The first mathematician to notice that Cauchy's proof could be performed on polyhedra with *bent faces* was Crelle ([1826–7], pp. 671–2), but he still carefully stuck to *straight edges*. For Cayley however it seemed recognisable '*at first sight*' that 'the theory would not be materially altered by allowing the edges to be curved lines' ([1861], p. 425). The same remark was made independently in Germany by Listing ([1861], p. 99) and in France by Jordan ([1866a], p. 39).

much more revolutionary than that: they *erase* the crucial naive concepts completely and *replace* them by proof-generated concepts.[1] The naive term 'polyhedron', even after being stretched by refutationists, denoted something that was crystal-like, a solid with 'plane' faces, straight edges. The proof-ideas swallowed this naive concept and fully digested it. In the different proof-generated theorems we have nothing of the naive concept. That disappeared without trace. Instead each proof yields its characteristic proof-generated concepts, which refer to stretchability, pumpability, photographability, projectability and the like. The old problem disappeared, new ones emerged. After Columbus one should not be surprised if *one does not solve the problem one has set out to solve.*

SIGMA: So the 'theory of solids', the original 'naive' realm of the Euler conjecture, dissolves, and the remodelled conjecture reappears in projective geometry if proved by Gergonne, in analytical topology if proved by Cauchy, in algebraic topology if proved by Poincaré...

PI: Quite right. And now you will understand why I formulated the theorems not, like Alpha or Beta, as: 'All Gergonne-polyhedra are Eulerian', 'All Cauchy-polyhedra are Eulerian', and so on, but rather as: 'All Gergonnian objects are Eulerian', 'All Cauchy objects are Eulerian', and so on.[2] *So I find it uninteresting to quarrel not only about the exactness of naive concepts but also about the truth or falsehood of naive conjectures.*

BETA: But surely we can retain the term 'polyhedron' for our favourite proof-generated term, say, 'Cauchy-objects'?

PI: If you like, but remember that *your term no longer denotes what it set out to denote*: that its naive meaning has disappeared and that now it is used...

BETA: ...for a more general, improved concept!

THETA: No! For a totally different, novel concept.

---

[1] *This theory of concept-formation weds concept-formation to proofs and refutations.* Pólya wed it to *observations*: 'When the physicists started to talk about "electricity", or the physicians about "contagion", these terms were vague, obscure, muddled. The terms that the scientists use today, such as "electric charge", "electric current", "fungus infection", "virus infection", are incomparably clearer and more definite. Yet what a tremendous amount of observation, how many ingenious experiments lie between the two terminologies, and some great discoveries too. Induction changed the terminology, clarified the concepts. We can illustrate also this aspect of the process, the inductive clarification of concepts, by suitable mathematical examples.' ([1954], vol. 1, p. 55.) But even this mistaken inductivist theory of concept-formation is preferable to the attempt to make concept-formation *autonomous*, to make 'clarification' or 'explication' of concepts a *preliminary* to any scientific discussion.

[2] See above, p. 66.

SIGMA: I think your views are paradoxical!

PI: If you mean by paradoxical 'an opinion not yet generally received',[1] and possibly inconsistent with some of your ingrained naive ideas, never mind: you only have to replace your naive ideas with the paradoxical ones. This may be a way to 'solve' paradoxes. But what particular view of mine do you have in mind?

SIGMA: You remember, we found that some star-polyhedra are Eulerian while some others are not. We were looking for a proof that would be deep enough to explain the Eulerianness both of ordinary and star-polyhedra...

EPSILON: I have it.[2]

SIGMA: I know. But just for the sake of argument let us imagine that there is no such proof, but that somebody offers, in addition to Cauchy's proof for Eulerian 'ordinary' polyhedra, a corresponding but altogether different proof for Eulerian star-polyhedra. Would you then, Pi, because of these two different proofs, propose to split into two what we formerly classified as one? And would you have two completely different things united under one name just because somebody finds a common explanation for some of their properties?

PI: Of course I would. I certainly wouldn't call a whale a fish, a radio a noisy box (as aborigines may do), and I am not upset when a physicist refers to glass as a liquid. Progress indeed replaces *naive classification* by *theoretical classification*, that is, by theory-generated (proof-generated, or if you like, explanation-generated) classification. Conjectures and concepts both have to pass through the purgatory of proofs and refutations. *Naive conjectures and naive concepts are superseded by improved conjectures (theorems) and concepts (proof-generated or theoretical concepts) growing out of the method of proofs and refutations.* And as theoretical ideas and concepts supersede naive ideas and concepts, theoretical language supersedes naive language.[3]

---

[1] Hobbes [1656], Animadversions upon the Bishop's Reply No. xxi.

[2] See above, p. 65, footnote 1.

[3] It is interesting to follow the gradual changes from the rather naive classification of polyhedra to the highly theoretical one. The first naive classification which covers not only simple polyhedra comes from Lhuilier: a classification according to the number of *cavities, tunnels* and '*inner polygons*' (see p. 79, footnote 1).

    (a) *Cavities.* Euler's first proof and, incidentally, Lhuilier's own ([1812–13a], pp. 174–7), rested on the decomposition of the *solid*, either by cutting off its corners one by one, or by decomposing it into pyramids from one or more points in the inside. Cauchy's proof-idea however – Lhuilier did not know about it – rested on the decomposition of the polyhedral *surface*. When the theory of polyhedral surfaces finally superseded the theory of polyhedral solids, cavities became uninteresting: *one* 'polyhedron with

OMEGA: In the end we shall arrive from naive, accidental, merely nominal classification to the final true, real, classification, to perfect language![1]

(c) Logical and heuristic refutations revisited

PI: Let me take up again some of the issues which have arisen in connection with deductive guessing. First let us take the problem of heuristic versus logical counterexamples as raised in the discussion between Alpha and Theta.

My exposition has shown, I think, that even the so-called 'logical' counterexamples were heuristic. In the originally intended interpretation there is no inconsistency between: (a) all polyhedra are Eulerian, and (b) the picture-frame is not Eulerian.

If we keep to the tacit semantical rules of our original language our counterexamples are not counterexamples. They are turned into

cavities' turns into a whole *class* of polyhedra. Thus our old monster-barring *Definition 2* (p. 14) became a proof-generated, theoretical definition, and the taxonomical concept of 'cavity' disappeared from the mainstream of growth.

(b) *Tunnels.* Already Listing pointed to the unsatisfactoriness of this concept (see p. 79, footnote 1). The replacement came not from any 'explication' of the 'vague' concept of tunnel, as a Carnapian might be tempted to expect, but from trying to prove and refute Lhuilier's naive conjecture about the Euler-characteristic of polyhedra with tunnels. In the course of this process the concept of polyhedron with n tunnels disappeared and proof-generated 'multiply-connectedness' (what we called 'n-spheroidness') took its place. In some papers we find the naive term retained for the new proof-generated concept: Hoppe defines the number of 'tunnels' by the number of cuts that leave the polyhedron connected ([1879], p. 102). For Ernst Steinitz the concept of tunnel is already so theory-impregnated that he is unable to find any 'essential' difference between Lhuilier's naive classification according to the number of tunnels and the proof-generated classification according to multiply-connectedness; therefore he regards Listing's criticism of Lhuilier's classification as 'largely unjustified' ([1914–31], p. 22).

(c) '*Inner polygons.*' This naive concept too was soon replaced first by ringshaped, then by multiply-connected, faces (also cf. p. 79, footnote 1), (*replaced*, not 'explicated', for 'ringshaped face' is surely not an explication of 'inner polygon'). When, however, the theory of polyhedral surfaces was superseded on the one hand by the topological theory of surfaces, and on the other hand by graph-theory, the problem of how multiply-connected faces influence the Euler-characteristic of a polyhedron lost all its interest.

Thus, out of the three key concepts of the first naive classification, only one was 'left', and even that in a hardly recognisable form – the generalised Euler formula was, for the moment, reduced to $V - E + F = 2 - 2n$. (For further developments cf. p. 89, footnote 3.)

[1] As far as naive classification is concerned, nominalists are close to the truth when claiming that the only thing that polyhedra have in common is their name. But after a few centuries of proofs and refutations, as the theory of polyhedra develops, and theoretical classification replaces naive classification, the balance changes in favour of the realist. The problem of universals ought to be reconsidered in view of the fact that, as knowledge grows, languages change.

logical counterexamples only by changing the rules of the language by concept-stretching.

GAMMA: Do you mean that *all* interesting refutations are heuristic?

PI: Exactly. You cannot separate refutations and proofs on the one hand and changes in the conceptual, taxonomical, linguistic framework on the other. Usually, when a 'counterexample' is presented, you have a choice: either you refuse to bother with it, since it is not a counter-example at all in your *given* language L_1, or you agree to change your language by concept-stretching and accept the counterexample in your new language L_2...

ZETA: ...and *explain* it in L_3!

PI: According to traditional static rationality you would have to make the first choice. Science teaches you to make the second.

GAMMA: That is, we may have two statements that are consistent in L_1, but we switch to L_2 in which they are inconsistent. Or, we may have two statements that are inconsistent in L_1, but we switch to L_2 in which they are consistent. As knowledge grows, languages change. 'Every period of creation is at the same time a period in which the language changes.'[1] The growth of knowledge cannot be modelled in any given language.

PI: That is right. Heuristic is concerned with language-dynamics, while logic is concerned with language-statics.

(d) Theoretical versus naive concept-stretching. Continuous versus critical growth

GAMMA: You promised to come back to the question of whether or not deductive guessing offers us a continuous pattern of the growth of knowledge.

PI: Let me first sketch some of the many *historical* forms which this *heuristic* pattern can take.

The *first main pattern* is when naive concept-stretching outstrips theory by far and produces a vast chaos of counterexamples: our naive

---

[1] Félix [1957], p. 10. According to logical positivists, the *exclusive* task of philosophy is to construct 'formalised' languages in which artificially congealed states of science are expressed (see our quotation from Carnap above, p. 1). But such investigations scarcely get under way before the rapid growth of science discards the old 'language system'. Science teaches us not to respect any given conceptual-linguistic framework lest it should turn into a conceptual prison – language analysts have a vested interest in at least slowing down this process, in order to justify their linguistic therapeutics, that is, to show that they have an all-important feedback to, and value for, science, that they are not de-generating into 'fairly dried-up petty-foggery' (Einstein [1953]). Similar criticisms of logical positivism have been made by Popper: see e.g. his [1959], p. 128, footnote *3.

concepts are loosened but no theoretical concepts replace them. In this case deductive guessing may catch up – piecemeal – with the backlog of counterexamples. This is, if you like, a continuous 'generalising' pattern – but do not forget that it starts with refutations, that its continuity is the piecemeal explanation by a growing theory of the heuristic refutations of its first version.

GAMMA: Or, 'continuous' growth only indicates that refutations are miles ahead!

PI: That is right. But it may happen that each single refutation or expansion of naive concepts is *immediately* followed by an expansion of the theory (and theoretical concepts) which explains the counterexample; 'continuity' then gives place to an exciting alternation of concept-stretching refutations and ever more powerful theories, of *naive concept-stretching* and explanatory *theoretical concept-stretching*.

SIGMA: Two accidental historical variations on the same heuristic theme!

PI: Well, there is not really much difference between them. In both of them *the power of the theory lies in its capacity to explain its refutations in the course of its growth*. But there is a *second main pattern* of deductive guessing...

SIGMA: Yet another accidental variation?

PI: Yes, if you like. In this variation however the growing theory not only *explains* but *produces* its refutations.

SIGMA: What?

PI: In this case theoretical growth overtakes – and, indeed, eliminates – naive concept-stretching. For example, one starts with, say, Cauchy's theorem, without a single counterexample on the horizon. Then one tests the theorem by transforming the polyhedron in all possible ways: cutting it into two, cutting off pyramidal corners, bending it, distorting it, pumping it up...Some of these test-ideas will lead to proof-ideas[1] (by arriving at something known to be true and then turning back, that is, by following the Pappian analysis-synthesis pattern), but some – like Zeta's 'double-pasting test' – will lead us, not back to something already known, but to real novelty, to some heuristic refutation of the tested proposition – *not by extending a naive concept, but by extending the theoretical framework*. This sort of refutation is self-explanatory...

IOTA: How dialectical! Tests turning into proofs, counterexamples that become examples by the very method of their construction...

---

[1] Pólya discriminates between 'simple' and 'severe' tests. 'Severe' tests may give 'the first hint of a proof' ([1954], vol. I, pp. 34–40).

Pi: Why dialectical? The test of one proposition turns into the proof of *another*, deeper proposition, counterexamples of the first into examples of the second. Why call confusion dialectic? But let me come back to my point: I do not think that my second main pattern of deductive guessing could be regarded – as Alpha would have it – as continuous growth of knowledge.

ALPHA: Of course it can. Compare our method with Omega's idea of replacing one proof-idea with a radically different, deeper one. Both methods increase content, but while in Omega's method one *replaces* operations of the proof that are applicable in a narrow domain by operations which are applicable in a wider domain, or, more radically, replaces the whole proof by one that is applicable in a wider domain – deductive guessing *extends* the given proof by adding operations which widen its applicability. Is this not continuity?

SIGMA: That is right! We deduce from the theorem a chain of ever wider theorems! From the special case ever more general cases! Generalisation by deduction![1]

Pi: But full of counterexamples, once you recognise that *any* increase of content, *any* deeper proof follows or generates heuristic refutations of the previous poorer theorems...

ALPHA: Theta expanded 'counterexample' to cover heuristic counterexamples. You now expand it to cover heuristic counterexamples that never actually exist. Your claim that your 'second pattern' is full of counterexamples is based on the expansion of the concept of counterexample to counterexamples with zero life-time, whose discovery coincides with their explanation! But why should all intellectual activity, every struggle for increased content in a unified theoretical framework, be 'critical'? Your dogmatic 'critical attitude' is obscuring the issue!

TEACHER: The issue between you and Pi is certainly obscure – for your 'continuous growth' and Pi's 'critical growth' are perfectly consistent. I am more interested in the *limitations*, if any, of deductive guessing, or 'continuous criticism'.

---

[1] In *informal* logic there is nothing wrong with the 'fact, so usual in mathematics and still so surprising to the beginner, or to the philosopher who takes himself for advanced, that the general case can be logically equivalent to a special case' (Pólya [1954], vol. 1, p. 17). Also cf. Poincaré [1902], pp. 31–3.

(e) The limits of the increase in content. Theoretical versus naive refutations

PI: I think that sooner or later 'continuous' growth is bound to reach a dead-end, a *saturation point* of the theory.

GAMMA: But surely I can always stretch some of the concepts!

PI: Of course. *Naive* concept-stretching may go on – but *theoretical* concept-stretching has limits! Refutations by naive concept-stretching are only gadflies that prod us to catch up by theoretical concept-stretching. So there are two sorts of refutations. We *stumble* on the first sort by coincidence or good fortune, or by an arbitrary expansion of some concept. They are like miracles, their 'anomalous' behaviour is unexplained; we accept them as *bona fide* counterexamples only because we are used to accepting concept-stretching criticism. I shall call these *naive* counterexamples or *freaks*. Then there are the *theoretical counterexamples*: these are either originally produced by proof-stretching or, alternatively, they are freaks which are reached by stretched proofs, explained by them, and thereby raised to the status of theoretical counterexamples. Freaks have to be looked upon with great suspicion: they may not be genuine counterexamples, but instances of a quite different theory – if not outright mistakes.

SIGMA: But what shall we do when we get stuck? When we cannot turn our naive counterexamples into theoretical ones by expanding our original proof?

PI: We may probe again and again whether or not our theory still has some hidden capacity for growth. Sometimes, however, we have good reason to give up. For instance, as Theta rightly pointed out, if our deductive guessing starts from a vertex we cannot very well ever expect it to explain the vertexless cylinder.

ALPHA: So after all, the cylinder was not a monster, but a freak!

THETA: But freaks should not be played down! They are the *real* refutations: they cannot be fitted into a pattern of continuous 'generalisations', and may actually force us to revolutionise our theoretical framework...[1]

---

[1] Cayley [1861] and Listing [1861] took the stretching of the basic concepts of the theory of polyhedra seriously. Cayley defined *edge* as 'the path from a summit to itself, or to any other summit' but allowed edges to degenerate into vertexless closed curves, which he called 'contours' (p. 426). Listing had one term for edges, whether with two, one, or no vertices: '*lines*' (p. 104). Both realised that a completely new theory was needed to explain the 'freaks' which they naturalised with their liberal conceptual framework – Cayley invented the '*Theory of Partitions of a Close*', Listing, one of the great pioneers of modern topology, the '*Census of Spatial Complexes*'.

OMEGA: Good! One may get to a *relative saturation point* of a *particular* chain of deductive guessing – but then one finds a revolutionary, new, deeper proof-idea that has more explanatory power. At the end one still gets to a *final* proof – without limit, without saturation point, without freaks to refute it!

PI: What? A single unified theory to explain *all* the phenomena of the universe? Never! Sooner or later we shall approach something like an *absolute saturation point*.

GAMMA: I don't really mind whether we do or not. If a counter-example can be explained by a cheap, *trivial* extension of the proof, I would already regard it as a freak. I repeat: I really do not see any point in generalising 'polyhedron' to include a polyhedron with cavities: this is not one polyhedron, but a class of polyhedra. I would also forget about 'multiply-connected faces' – why not draw the missing diagonals? As to the generalisation that includes twintetra-hedra, I would reach for my gun: it serves only for making up complicated, pretentious formulas for nothing.

RHO: At last you rediscover my method of monster-adjustment![1] It relieves you of shallow generalisation. Omega should not have called content 'depth'; *not every increase in content is also an increase in depth*: think of (6) and (7)![2]

ALPHA: So you would stop at (5) in my series?

[1] See above, pp. 30–3 and 38–9.

[2] Quite a few mathematicians cannot distinguish the trivial from the non-trivial. This is especially awkward when a lack of feeling for relevance is coupled with the illusion that one can construct a *perfectly complete* formula that covers all conceivable cases (cf. p. 80, footnote 1). Such mathematicians may work for years on the 'ultimate' generalisation of a formula, and end up by extending it with a few trivial corrections. The excellent mathematician, J. C. Becker, provides an amusing example: after many years' work he produced the formula $V - E + F = 4 - 2n + q$ where n is the number of cuts that is needed to divide the polyhedral surface into simply-connected surfaces for which $V - E + F = 1$, and q is the number of diagonals that one has to add to reduce all the faces to simply-connected ones ([1869a], p. 72). He was very proud of his achievement, which – he claimed – shed 'completely new light', and even 'brought to a conclusion' 'a subject in which people like Descartes, Euler, Cauchy, Gergonne, Legendre, Grunert, and von Staudt, took interest' before him (p. 65). But three names were missing from his reading list: Lhuilier, Jordan and Listing. When he was told about Lhuilier, he published a sad note, admitting that Lhuilier knew all this more than fifty years before. As for Jordan, he was not interested in ringshaped faces, but happened to take an interest in open polyhedra with boundaries, so that in his formula m, the number of boundaries, figures in addition to n ([1866b], p. 86). So Becker – in a new paper [1896b] – combined Lhuilier's and Jordan's formulas into $V - E + F = 2 - 2n + q + m$ (p. 343). But in his embarrassment he was too hasty, and had not digested Listing's long paper. So he sadly concluded his [1869b] with 'Listing's generalisation is still wider'. (By the way, he later tried to extend his formula also to star-polyhedra ([1874]; cf. above, p. 31, footnote 4.)

GAMMA: Yes. (6) and (7) are not growth, but degeneration! Instead of going on to (6) and (7), I would rather find and explain some *exciting* new counterexample![1]

ALPHA: You may be right after all. But who decides *where* to stop? Depth is only a matter of taste.

GAMMA: Why not have mathematical critics just as you have literary critics, to develop mathematical taste by public criticism? We may even stem the tide of pretentious trivialities in mathematical literature.[2]

SIGMA: If you stop at (5) and turn the theory of polyhedra into a theory of triangulated spheres with n handles, how can you, if the need arises, deal with trivial anomalies like those explained in (6) and (7)?

MU: Child's play!

THETA: Right. Then we stop at (5) for the moment. *But can we stop?* Concept-stretching may refute (5)! We may ignore the stretching of a concept if it yields a counterexample that shows up the poverty of the content of our theorem. But if the stretching yields a counterexample that shows up its plain falsehood, what then? We may refuse to apply our content-increasing *Rule 4* or *Rule 5* to explain a freak, but we have to apply our content-preserving *Rule 2* to ward off refutation by a freak.

GAMMA: That is it! We may dismiss cheap '*generalisations*', but we can hardly dismiss 'cheap' *refutations*.

SIGMA: Why not build up a monster-barring definition of 'polyhedron', adding a new clause for each freak?

THETA: In both cases our old nightmare, vicious infinity, is back again.

ALPHA: While you are increasing content, you develop ideas, do

---

[1] Some people may entertain philistine ideas about *a law of diminishing returns in refutations*. Gamma, for one, certainly does not. We shall not now discuss *one-sided* polyhedra (Möbius, [1865]) or *n-dimensional* polyhedra (Schläfli, [1852]). These would confirm Gamma's expectation that totally unexpected concept-stretching refutations may always give the whole theory a new – possibly revolutionary – push.

[2] Pólya points out that shallow, cheap, generalisation is 'more fashionable nowadays than it was formerly. It dilutes a little idea with a big terminology. The author usually prefers to take even that little idea from somebody else, refrains from adding any original observation, and avoids solving any problem except a few problems arising from the difficulties of his own terminology. It would be very easy to quote examples, but I don't want to antagonize people' ([1954], vol. I, p. 30). Another of the greatest mathematicians of our century, John von Neumann, also warned against this 'danger of degeneration', but thought it would not be so bad 'if the discipline is under the influence of men with an exceptionally well-developed taste' ([1947], p. 196). One wonders, though, whether the 'influence of men with an exceptionally well-developed taste' will be enough to save mathematics in our 'publish or perish' age.

mathematics; after it you clarify concepts, do linguistics. Why not stop altogether when one stops increasing content? Why be trapped in vicious infinities?

MU: Not mathematics versus linguistics again! Knowledge never profits from such disputes.

GAMMA: The term 'never' soon turns into 'soon'. I am all for taking up our old discussion again.

MU: But we already ended up in a deadlock! Or does anybody have anything new to say?

KAPPA: I think I have.

9. How Criticism may turn Mathematical Truth into Logical Truth

(a) Unlimited concept-stretching destroys meaning and truth

KAPPA: Alpha already said that our 'old method' leads to vicious infinity.[1] Gamma and Lambda answered with the hope that the stream of refutations might peter out:[2] but now that we understand the mechanism of refutational success – concept-stretching – we know that theirs was a vain hope. For any proposition there is always some sufficiently narrow interpretation of its terms, such that it turns out true, and some sufficiently wide interpretation such that it turns out false. Which interpretation is intended and which unintended depends of course on our intentions. The first interpretation may be called the *dogmatist, verificationist or justificationist interpretation*, the second the *sceptical, critical or refutationist interpretation*. Alpha called the first a conventionalist stratagem[3] – but now we see that the second is one too. You all ridiculed Delta's dogmatist interpretations of the naive conjecture[4] and then Alpha's dogmatist interpretation of the theorem.[5] But concept-stretching will refute *any* statement, and will leave no true statement whatsoever.

GAMMA: Wait. True, we stretched 'polyhedron' – then tore it up and threw it away: as Pi pointed out, the naive concept 'polyhedron' does not figure in the theorem any more.

KAPPA: But then you will start stretching a term in the theorem – a theoretical term, won't you? You yourself chose to stretch 'simply-connected face' to include the circle and the jacket of the cylinder.[6] You

---

[1] See above, p. 53.   [2] See above, ibid.
[3] Alpha in fact did not use this Popperian term explicitly; see above, p. 21.
[4] See above, §4,(b).   [5] See above, §5.   [6] See above, pp. 42–6.

implied that it was a matter of intellectual honesty to stick one's neck out, to achieve the respectable status of refutability, i.e. to make the refutationist interpretation possible. But because of concept-stretching, refutability means refutation. So you slide on to the infinite slope, refuting *each* theorem and replacing it by a more 'rigorous' one – by one whose falsehood has not been 'exposed' yet! But *you never get out of falsehood.*

SIGMA: What if we stop at a certain point, adopt justificationist interpretations, and don't budge either from the truth or from the particular linguistic form in which that truth was expressed?

KAPPA: Then you will have to ward off concept-stretching counterexamples with monster-barring definitions. Thus you will slide on to another infinite slope: you will be forced to admit of each 'particular linguistic form' of your true theorem that it was not precise enough, and you will be forced to incorporate in it more and more 'rigorous' definitions couched in terms whose vagueness has not been exposed yet! But *you never get out of vagueness.*[1]

THETA [*aside*]: What is wrong with a heuristic where vagueness is the price we pay for growth?

ALPHA: I told you: precise concepts and unshakable truths do not dwell in language, but only in thought!

GAMMA: Let me challenge you, Kappa. Take the theorem as it stood, after we took account of the cylinder: 'For all simple objects with simply-connected faces such that the edges of the faces terminate in vertices, $V - E + F = 2$.' How would you refute *this* by the method of concept-stretching?

KAPPA: First I go back to the defining terms and spell out the proposition in full. Then I decide which concept to stretch. For instance, 'simple' stands for 'stretchable onto a plane after having had a face removed'. I shall stretch 'stretching'. Take the already discussed twin-tetrahedra – the pair with an *edge* in common (fig. 6(*a*)). It is simple, its faces are simply-connected, but $V - E + F = 3$. So our theorem is false.

GAMMA: But this twintetrahedron is *not* simple!

KAPPA: Of course it is simple. Removing any face, I can stretch it

---

[1] *Editors' note:* Kappa's claim that vagueness is inescapable is correct (*some* terms are bound to be primitive). But he is wrong to think that this means that one can always produce counterexamples by 'concept-stretching'. By definition, a valid proof is one in which, *no matter how one interprets the descriptive terms*, one never produces a counterexample – i.e. its validity does not depend on the meaning of the descriptive terms, which can thus be stretched however one likes. This is pointed out by Lakatos himself below, p. 103 and (more clearly), chapter 2, p. 124.

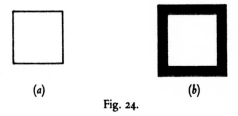

(a)   (b)

Fig. 24.

on to a plane. I just have to be careful, when I get to the critical edge, that I do not tear anything there when opening the second tetrahedron along that edge.

GAMMA: But this is not stretching! You *tear* – or *split* – the edge into two edges! You certainly cannot map one point onto two: *stretching is a bicontinuous one-one mapping*!

KAPPA: *Def. 7*? I am afraid this narrow, dogmatist interpretation of 'stretching' does not appeal to *my* common sense. For instance, I can well imagine stretching a square (fig. 24(*a*)) into two nested squares by stretching the boundary lines (fig. 24(*b*)). Would you call this stretch a tear or a split, just because it is not a 'bicontinuous one-one mapping'? By the way, I wonder why you did not define stretching as a transformation that leaves V, E and F unaltered, and have done with it?

GAMMA: Right, you win again. I either have to agree to your refutationist interpretation of 'stretching' and expand my proof, or find a deeper one, or incorporate a lemma – or I have to introduce a new monster-barring definition. Yet in any of these cases I shall always make my defining terms clearer and clearer. Why should I not arrive at a point where the meanings of the terms will be so crystal clear that there will only be one single interpretation, as is the case with $2+2 = 4$? There is nothing elastic about the meaning of these terms and nothing refutable about the truth of this proposition, which shines for ever in the natural light of reason.

KAPPA: Dim light!

GAMMA: Stretch, if you can.

KAPPA: But this is child's play! In certain cases two and two make five. Suppose we ask for the delivery of two articles each weighing two pounds; they are delivered in a box weighing one pound; then in this package two pounds and two pounds will make five pounds!

GAMMA: But you get five pounds by adding *three* weights, 2 and 2 and 1!

KAPPA: True, our operation '2 and 2 make 5' is not an addition in the originally intended sense. But we can make the result hold true by a simple stretching of the meaning of addition. Naive addition is a very special case of packing where the weight of the covering material is zero. We have to build this lemma into the conjecture as a condition: our improved conjecture will be: '2 + 2 = 4 for "weightless" addition'.[1] The whole story of algebra is a series of such concept- and proof-stretchings.

GAMMA: I think you take 'stretching' a bit far. Next time you will interpret 'plus' as 'times' and consider it a refutation! Or you will interpret 'all' as 'no' in 'All polyhedra are polyhedra'! You stretch the concept of concept-stretching! We have to demarcate refutation by *rational stretching* from 'refutation' by *irrational stretching*. We cannot allow you to stretch any term you like just as you like.

We must pin down the concept of counterexample in crystal-clear terms!

DELTA: Even Gamma has turned into a monster-barrer: now he wants a monster-barring definition of concept-stretching refutation. *Rationality, after all, depends on inelastic, exact, concepts!*[2]

KAPPA: *But there are no such concepts! Why not accept that our ability to specify what we mean is nil, therefore our ability to prove is nil? If you want mathematics to be meaningful, you must resign of certainty. If you want certainty, get rid of meaning. You cannot have both. Gibberish is safe from refutations, meaningful propositions are refutable by concept-stretching.*

GAMMA: Then your last statements can also be refuted – and you know it. 'Sceptics are not a sect of people who are persuaded of what they say, but a sect of liars.'[3]

KAPPA: Swear-words: the last resort of reason!

(b) Mitigated concept-stretching may turn mathematical truth into logical truth

THETA: I think Gamma is right about the need for demarcating rational from irrational concept-stretching. For concept-stretching has come a long way, and has changed from a mild, rational activity to a radical, irrational one.

Originally, criticism concentrates exclusively on the *slight* stretching of *one particular* concept. It has to be *slight*, so that we do not notice it;

---

[1] Cf Félix [1957], p. 9.
[2] Gamma's demand for a crystal-clear definition of 'counterexample' amounts to a demand for crystal-clear, inelastic concepts in the metalanguage as a condition of rational discussion.   [3] Arnauld and Nicole [1724], pp. xx–xxi.

if its real – stretching – nature were discovered, it might not be accepted as legitimate criticism. It concentrates on *one particular* concept, as in the case of our rather unsophisticated universal propositions: '*All A's are B's*'. Criticism then means finding a slightly stretched *A* (in our case *polyhedron*) that is not *B* (in our case *Eulerian*).

But Kappa sharpened this in two directions. First, to submit *more than one* constituent of the proposition under attack to concept-stretching criticism. Second, to turn concept-stretching from a surreptitious and rather modest activity into *open deformation* of the concept, like the deformation of 'all' into 'no'. Here any meaningful translation of the terms under attack that renders the theorem false is accepted as refutation. I would then say that *if a proposition cannot be refuted with respect to the constituents a, b, ..., then it is logically true with respect to these constituents.*[1] Such a proposition is the end-result of a long critical-speculative process in the course of which the meaning-load of some terms is completely transferred to the remaining terms and to the form of the theorem.

Now all that Kappa says is that there are no propositions which are logically true with respect to *all* their constituents. But there may be logically true propositions with respect to *some* constituents, so that the stream of refutations can only be opened up again if new stretchable constituents are added. If we go the whole hog, we end up in irrationalism – but we need not. Now where should we draw the borderline? We may very well allow concept-stretching only for a distinguished subset of constituents which become the prime targets of criticism. Logical truth will not depend on their meaning.

SIGMA: So after all we took Kappa's point: we made truth independent of the meaning of at least *some* of the terms!

THETA: That is right. But if we want to defeat Kappa's scepticism, and escape his vicious infinities, we certainly have to stop concept-stretching at the point where it ceases to be a tool of growth and becomes a tool of destruction: we may have to find out which are those terms whose meaning can be stretched only at the cost of destroying the basic principles of rationality.[2]

---

[1] This is a slightly paraphrased version of Bolzano's definition of logical truth ([1837], § 147). Why Bolzano, in the 1830s, proposed his definition, is a puzzling question, especially since his work anticipates the concept of model, one of the greatest innovations in nineteenth-century mathematical philosophy.

[2] Nineteenth-century mathematical criticism stretched more and more concepts, and shifted the meaning-load of more and more terms onto the *logical form* of the propositions and onto the meaning of the few (as yet) unstretched terms. In the 1930s this process seemed to slow down and the demarcation line between unstretchable ('logical')

KAPPA: Can we stretch the concepts in your theory of critical rationality? Or will that be manifestly true, formulated in unstretchable, exact terms which do not need to be defined? Will your theory of criticism end in a 'retreat to commitment': is everything criticisable except for your theory of criticism, your 'metatheory'?[1]

OMEGA [*to Epsilon*]: I do not like this shift from Truth to rationality. *Whose* rationality? I sense conventionalist infiltration.

BETA: What are you talking about? I understand Theta's 'mild pattern' of concept-stretching. I also understand that concept-stretching may attack more than one term: we saw this when Kappa stretched 'stretching' or when Gamma stretched 'all'...

SIGMA: Surely Gamma stretched 'simply-connected'!

BETA: But no. 'Simply-connected' is an abbreviation – he only stretched the term 'all' that occurred among the defining terms.[2]

THETA: Come back to the point. You are unhappy about 'open', radical concept-stretching?

BETA: Yes. Nobody would accept this last brand as genuine refutation! I quite see that the mild concept-stretching trend of heuristic criticism that Pi uncovered is a most important vehicle of mathematical growth. But mathematicians will never accept this last, wild form of refutation!

TEACHER: You are wrong, Beta. They *did* accept it, and their acceptance was a turning point in the history of mathematics. *This revolution in mathematical criticism changed the concept of mathematical truth, changed the standards of mathematical proof, changed the patterns of mathematical*

terms and stretchable ('descriptive') terms seemed to become stable. A list, containing a small number of logical terms, came to be widely agreed upon, so that a general definition of logical truth became possible; logical truth was no longer 'with respect to' an *ad hoc* list of constituents. (Cf. Tarski [1935].) Tarski was, however, puzzled about this demarcation and wondered whether, after all, he would have to return to a relativised concept of counterexample, and consequently, of logical truth (p. 420) – like Bolzano's, of which, by the way, Tarski did not know. The most interesting result in this direction was Popper's [1947–8] from which it follows that one cannot give up further logical constants without giving up some basic principles of rational discussion.

[1] 'Retreat to commitment' is Bartley's expression [1962]. He investigates the problem of whether a rational defence of critical rationalism is possible mainly with respect to *religious* knowledge – but the problem-patterns are very much the same with respect to *mathematical* knowledge.

[2] See above, pp. 42–6. Gamma did, in fact, want to remove some meaning-load from 'all', so that it no longer applied only to non-empty classes. The modest stretching of 'all' by removing 'existential import' from its meaning and thereby turning the empty set from a monster into an ordinary *bourgeois* set was an important event – connected not only with the Boolean set-theoretical re-interpretation of Aristotelian logic, but also with the emergence of the concept of vacuous satisfaction in mathematical discussion.

growth![1] But now let us close our discussion for the time being: we shall discuss this new stage some other time.

SIGMA: But then nothing is settled. We can't stop *now*.

TEACHER: I sympathise. This latest stage will have important feed-backs to our discussion.[2] But a scientific inquiry 'begins and ends with problems'.[3] [*Leaves the classroom.*]

BETA: But I had no problems at the beginning! And now I have nothing *but* problems!

[1] The concepts of criticism, counterexample, consequence, truth, and proof are in-separable; when they change, *the primary change occurs in the concept of criticism* and changes in the others follow.

[2] Cf. Lakatos [1962].   [3] Popper [1963*b*], p. 968.

2

Editors' Introduction

Poincaré's proof of the Descartes–Euler conjecture is referred to above.[1] In his doctoral thesis Lakatos introduced detailed consideration of this proof by a discussion of the arguments for and against the 'Euclidean' approach to mathematics. Parts of this discussion were incorporated by Lakatos into chapter 1 (see, e.g., pp. 50–6) and others were rewritten as parts of 'Infinite Regress and the Foundations of Mathematics' (Lakatos [1962]). We therefore omit this introductory discussion here.

The advocate of the Euclidean programme – the attempt to supply mathematics with indubitably true axioms couched in perfectly clear terms – has been Epsilon. Epsilon's philosophy is challenged, but the Teacher remarks that the most obvious and direct way to challenge Epsilon is to ask him to produce a proof of the Descartes–Euler conjecture which satisfies Euclidean standards. Epsilon takes up the challenge.

1. Translation of the Conjecture into the 'Perfectly Known' Terms of Vector-Algebra. The Problem of Translation

EPSILON: I accept the challenge. I shall prove that all simply-connected polyhedra with simply-connected faces are Eulerian.

TEACHER: Yes, I stated this theorem in a previous lesson.[2]

EPSILON: As I have pointed out, I first have to find the truth in order to prove it. Now I have nothing against using your method of proofs and refutations as a method of discovering the truth, but where you stop, I start. Where you stop improving, I start proving.[3]

ALPHA: But this long theorem is full of stretchable concepts. I do not think we shall find it difficult to refute it.

EPSILON: You will find it impossible to refute it. I shall pin down the meaning of each single term.

TEACHER: Go on.

EPSILON: First I shall use only the clearest possible concepts. Maybe sometime we shall be able to extend our perfect knowledge to cover

---

[1] See pp. 65 and 90.   [2] See above, p. 36.
[3] Epsilon is probably the first-ever Euclidean to appreciate the heuristic value of the proof-procedure. Until the seventeenth century, Euclideans approved the Platonic method of analysis as the method of heuristic; later they replaced it by the stroke of luck and/or genius.

optical cameras, paper and scissors, rubber balls and pumps, but *now* we should forget these things. *Finality* certainly cannot be reached by using all these various tools. Our previous failures, in my view, are rooted in the fact that we used methods which are alien to the simple, naked nature of polyhedra. The exuberant imagination which mobilised all these tools is completely mis-directed. It adduced external, alien, contingent elements which do not pertain to the essence of polyhedra and so no wonder it fails for some polyhedra. In order to get a perfect proof one has to restrict the range of tools used.[1] This is because this exuberant imagination makes *certainty* too difficult to attain. The truth of lemmas which hinge on the properties of rubber, lenses and so on, is difficult to guarantee. We should abandon scissors, pumps, cameras and the like, because 'for the understanding of a question we must abstract it from all that is superfluous, rendering it as simple as possible'.[2] I purge my theorem[3] and my proof of all these, and restrict them to the simplest and easiest things:[4] namely to vertices, edges and faces. I shall not define *these* terms as there cannot possibly be a disagreement about their meaning. I shall define any term which is in the least obscure in perfectly known 'primitive' terms.[5]

Now it is clear that none of the specific lemmas in any of the proofs was evidently true; they were just conjectures such as 'All polyhedra are pumpable into a ball' and so on. But now 'I require that no conjectures of any kind be allowed into the judgments we pass on the truth of things'.[6] I shall decompose the conjecture into lemmas which are not conjectures any longer but 'intuitions', that is, 'nondubious apprehensions of a pure and attentive mind which are born in the sole light of reason'.[7] Examples of these 'intuitions' are: *all polyhedra have faces*; *all faces have edges*; *all edges have vertices*. I shall not raise such questions as whether a polyhedron is a solid or a surface. These are

---

[1] In proof-analysis there is no limitation on the 'tools'. We can use any lemma, any concept. This is true of any growing, informal theory, where problem-solving is a catch-as-catch-can affair. In a formalised theory the tools are completely prescribed in the syntax of the theory. In the ideal case (where there is a decision procedure) problem-solving here is a ritual.

[2] These are Descartes's words in his [1628], Rule XIII.

[3] One should not forget that while proof-analysis *concludes* with a theorem, the Euclidean proof *starts* with it. In the Euclidean methodology there are no conjectures, only theorems.

[4] Descartes [1628], Rule IX.

[5] Pascal's rules for definitions ([1659], pp. 596–7): 'Not to define any given term which is perfectly known. Not to allow without definition any term in the least obscure or equivocal. To employ in the definition of terms only perfectly known or already explained words.'

[6] Descartes [1628], notes to Rule III. [7] Ibid.

vague notions and anyway superfluous for our purpose. For me a polyhedron consists of three sets: the set of V vertices (I shall call them $P_1^0, P_2^0, \ldots, P_V^0$), the set of E edges (I shall call them $P_1^1, P_2^1, \ldots, P_E^1$), and the set of F faces (I shall call them $P_1^2, P_2^2, \ldots, P_F^2$). In order to characterise a polyhedron we also need some sort of table that tells us which vertices belong to which edges, and which edges belong to which faces. I shall call these tables 'incidence matrices'.

GAMMA: I am a bit puzzled by your definition of polyhedra. In the first place, as you bother to define the notion of a polyhedron at all, I conclude that you do not consider it to be perfectly well known. But then where do you take your definition from? You defined the obscure concept of polyhedron in terms of the 'perfectly known' concepts of faces, edges and vertices. But your definition – namely that the polyhedron is a set of vertices, plus a set of edges, plus a set of faces, plus an incidence matrix, obviously fails to capture the intuitive notion of a polyhedron. It implies, for instance, that any polygon is a polyhedron, as is, say, a polygon with a free edge standing out of it. Now you must choose between two courses. You may say that 'the mathematician is not concerned with the current meaning of his technical terms... The mathematical definition creates the mathematical meaning'.[1] In this case to define the notion of a polyhedron is to drop the old notion altogether and to replace it by a new concept. But then any resemblance between your 'polyhedron' and any genuine polyhedron is entirely accidental, and you will not get any certain knowledge about genuine polyhedra by studying your mock-polyhedra. The other course is to stick to the idea that definition is clarification, that it makes essential features explicit, that it is a translation or a meaning-preserving transformation of a term into a clearer language. In this case your definitions are conjectures, they may be true, they may be false. How can you have a certainly true translation of a vague term into precise ones?

EPSILON: I admit you have taken me by surprise by this criticism. I thought you might doubt the absolute truth of my axioms, I thought you might ask how such *a priori* synthetic judgments are possible, and I prepared some counter-arguments, but I did not expect an attack on the line of definitions. But I suppose my answer is: I get my definitions, just as I get my axioms, by intuition. They are really of equal standing: you can take my definitions as additional axioms[2] or you can take my

[1] Pólya [1945], pp. 81–2.
[2] 'Definition as an undemonstrable statement of essential nature' (Aristotle, *Analytica Posteriora*, 94a).

axioms as implicit definitions.[1] They give the essence of the terms in questions.

TEACHER: Enough of philosophy! Let us see the proof. I do not like your philosophy, but I still may like your proof.

EPSILON: All right. I shall first translate the theorem to be proved into my perfectly simple and clear conceptual framework. My specific undefined terms will be: vertices, edges, faces and polyhedra. I shall sometimes refer to them as zero, one, two, and three dimensional polytopes,[2] or briefly, 0-polytopes, 1-polytopes, 2-polytopes and 3-polytopes.

ALPHA: But only ten minutes ago you defined polyhedra in terms of vertices, edges and faces!

EPSILON: I was wrong. That 'definition' was a stupid anticipation. I jumped to my judgment in a silly rush. True intuition, true interpretation, ripens slowly, and purging one's soul of conjectures takes time.[3]

BETA: You mentioned a moment ago some of your axioms, like: faces *have* edges, or to each face *belong* edges – 'belong to': is this another primitive term?

EPSILON: No. I register only terms *specific* to the theory in question, in this case the theory of polyhedra, but not the logical, set-theoretical, arithmetical ones of the underlying theory, with which I assume perfect familiarity. But let me now go on to the term 'simply-connected', which is certainly not absolutely clear. I shall define first simply-connectedness of polyhedra and then simply-connectedness of faces. I take simply-connectedness of polyhedra first. It is in fact the abbreviation of a long expression: a polyhedron is said to be simply connected, (1) if all closed loopless systems of edges have an inside and outside, and (2) if there is only one closed loopless system of faces – that which separates the inside from the outside of the polyhedron. Now this is full of rather vague terms, like 'closed', 'inside', 'outside' and so on. But I shall define all of them in perfectly known terms.

GAMMA: You have exorcised mechanical terms – like pumping,

---

[1] Gergonne [1818].

[2] That these terms can be subsumed under one single general abstract term was discovered by Schläfli ([1852]). He called them 'polyschemes'. Listing [1861] calls them 'Curian'. But it was Schläfli who extended the generalisation to more than three dimensions.

[3] 'The conclusions of human reason as ordinarily applied in matters of nature, I call for the sake of distinction, *Anticipations of Nature* (as a thing rash or premature). That which reason elicited from the facts by a just and methodical process, I call *Interpretation of Nature*' (Bacon [1620], XXVI).

cutting – as unreliable; now you jettison geometrical terms – like closedness. I think you are overdoing your purging zeal. 'A closed system of edges' is a perfectly clear concept, it need not be defined.

EPSILON: No, you are wrong. Would you call a star-polygon a closed system of edges? Maybe you would, because it has no loose end. But it does not 'enclose' any well defined area, and some may mean by a 'closed system of edges' a system of edges which does. So you have to make up your mind in one way or the other, and say in which way you have decided.

GAMMA: A star-polygon may not be bounded, but it is obviously *closed*.

EPSILON: I think that it is closed and that it is bounded too. The disagreement is already telling, but I shall produce some further evidence. I wonder whether or not you would say that the heptahedron is a closed system of faces and that it is bounded?

GAMMA: I have never heard of your heptahedron.

EPSILON: It is a rather interesting sort of polyhedron, as it is one-sided. There is no geometrical solid which it encloses, it does not separate the space into two parts, into an inside and an outside. Alpha, for instance, guided by his 'clear' geometrical intuition, said earlier that a closed system of faces bounds 'if it is the boundary between the inside of the polyhedron and the outside of the polyhedron'. I wonder whether he would say that the surface of the heptahedron does not bound? Or will getting acquainted with the heptahedron change his concept of 'bounding' systems? In this case I most humbly ask you: can *perfectly* known concepts be changed by experience? They cannot. Therefore 'closed', 'bounded' are not perfectly well known. Therefore I am going to define them.

THETA: Draw that heptahedron. I wonder what is it like?

EPSILON: All right. I start first with an ordinary familiar octahedron (see fig. 25). Now I add three squares in the planes spanned by the diagonals, for instance $A B C D$ (fig. 26).

DELTA: I should expect from a decent polyhedron that at the edges only two faces should meet. Here we have three.

EPSILON: Wait. I remove now four triangles in order to comply with this requirement: from the first half of the figure I remove the upper left-hand triangle and the lower right-hand triangle. From the part at the rear of the figure I remove the lower left-hand triangle and the upper right-hand triangle. Then only the four triangles shaded in the diagram remain (fig. 27). Thus we have obtained a figure consisting of

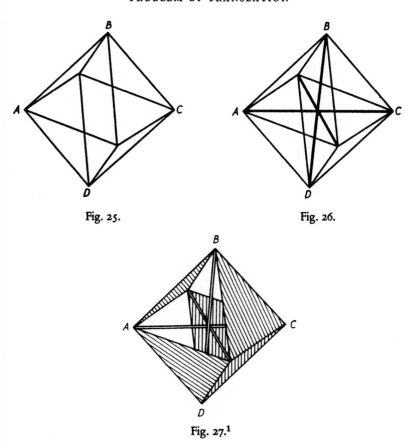

Fig. 25.

Fig. 26.

Fig. 27.[1]

four triangles and three squares. This is the heptahedron.[2] Its edges and vertices are the original edges and vertices of the octahedron. The diagonals of the octahedron are not edges of our figure but are lines in which it intersects itself. I do not attach much importance to geometrical intuition, I am not very interested in the fact that my polyhedron happens to be so uncomfortably embedded into three-dimensional space. This fact is not displayed by the incidence-matrices of my heptahedron. (By the way, the heptahedron can be embedded nicely without self-intersection into five-dimensional space.)[3]

---

[1] Figure 27 is redrawn from Hilbert and Cohn-Vossen [1932].
[2] Discovered by C. Reinhardt (see his [1885], p. 114).
[3] That one-sidedness or two-sidedness is dependent on the number of the dimensions of the space was first noticed by W. Dyck. See his [1888], p. 474.

Now does the surface of the heptahedron bound? The answer is 'no' if you define a surface as 'bounding' if and only if it is the boundary of the polyhedron in the sense that it separates the inside and the outside of the polyhedron in question. On the other hand, the answer is 'yes' if you define a surface as 'bounding' if and only if it is the boundary of the polyhedron in the sense that it contains all its faces. You see, you have to *define* 'bound', you have to define 'boundary'. These concepts may seem to have a touch of familiarity before one starts investigating the richness of polyhedral forms, but during such an investigation the original rough concepts split up and display a fine structure, and so you have to define your concepts carefully so that it is clear in which sense you are using them.

KAPPA: And then you have to put a veto on further investigation in order to avoid further splittings!

TEACHER: Epsilon, do not listen to Kappa. Refutations, inconsistencies, criticism in general are very important, but only if they lead to improvement. *A mere refutation is no victory.* If mere criticism, even though correct, had authority, Berkeley would have stopped the development of mathematics and Dirac could not have found an editor for his papers.

EPSILON: Do not worry, I dismissed Kappa's pointless heckling at once. I am now going on to define my terms, to translate everything into my few specific primitive terms – polytopes and incidence-matrices. I shall start by defining 'boundary'. The boundary of a k-polytope is the sum of the $(k\text{-}1)$ polytopes which belong to it according to the incidence-matrices. I shall call a sum of k-polytopes a k-chain. For instance the 'surface' of a polyhedron (or any part of it) is essentially a 2-chain. I define the boundary of a k-chain as the sum of the $(k-1)$ polytopes which belong to the k-chain, but instead of ordinary sum I take the sum *modulo* 2. This means that the following will hold:

$$0+0 = 0, \; 1+0 = 1, \; 0+1 = 1, \; 1+1 = 0.$$

You have to see that this is the *true definition* of the boundary of a k-chain.

BETA: Stop for a moment. I cannot easily follow your k-dimensional definitions. Let me think loudly about an example.[1] For instance the boundary of a *face* is, according to your definition, the set of edges which belong to it. Now when I join two faces, the common boundary

---

[1] *Editor's note:* 'Thinking loudly' was a technical term of Lakatosian English.

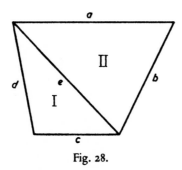

Fig. 28.

will not contain the edges which they both contain. So when adding the edges I shall omit those which occur in couples. For instance I take two triangles (fig. 28). The boundary of the first is $c+d+e$, the boundary of the second, $a+b+e$, the boundary of their join $a+b+e+c+d+e = a+b+c+d$. I see now why you introduced the *mod* 2 sums in your definition. Please carry on.

EPSILON: After having defined 'boundary' in perfectly known specific terms I shall now define 'closedness'. Hitherto either you had to rely on a vague insight, or you had to define closedness in each case separately: first the closedness of systems of edges, then the closedness of systems of faces. Now I show you that there is a general concept of closedness, applicable to any k-chain, independently of k. I shall call a k-chain a closed k-chain, or briefly, a k-circuit, if and only if its boundary is zero.

BETA: Stop for a moment. Let me see: an ordinary polygon is intuitively closed, and it is in fact closed according to your definition since its boundary is zero, as each vertex occurs twice in the boundary, and that makes zero in your *mod* 2 algebra. An ordinary simple polyhedron is closed, and again its boundary is zero, as in its boundary each edge occurs twice.

KAPPA [*aside*]: Beta certainly has to struggle to verify Epsilon's 'obvious and immediate insights'!

EPSILON: The next term to be elucidated is 'bound'. I shall say that a k-circuit bounds, if it is the boundary of a $(k+1)$-chain. For instance, the 'equator' of a spheroid polyhedron bounds, but the 'equator' of a toroid polyhedron does not. In this latter case the alternative idea, namely that it bounds the 'whole' of the polyhedron, is now ruled out, as the boundary of the whole of the polyhedron is empty. Now it is absolutely clear that for instance the heptahedron bounds.

BETA: You are a bit quick, but you seem to be right.

GAMMA: Can you prove that any bounding k-chain is a circuit? You defined 'bounding' only for circuits – you could have done it in general for chains. I suppose the reason for your restricted definition is this latent theorem.

EPSILON: That is right. I can prove that.

GAMMA: Another query. Some chains are circuits, some circuits bound. This seems to me to be in order. But I think that the boundary of a decent k-chain should be closed. For instance I could not possibly accept as a polyhedron a cube with the top missing; and I could not possibly accept as a polygon a square with an edge missing. Can you prove, that the boundary of any k-chain is closed?

EPSILON: Can I prove that the boundary of the boundary of any k-chain is zero?

GAMMA: That is it.

EPSILON: No, I cannot. This is indubitably true. It is an axiom. There is no need to prove it.

TEACHER: Go on, go on! I assume now you can translate our theorem into your perfectly known terms.

EPSILON: Yes. In brief, the translated theorem is: '*All polyhedra, all of whose circuits bound, are Eulerian*'. The specific term 'polyhedron' is undefined; I have already defined 'circuit' and 'bound' in perfectly known terms.

GAMMA: You have forgotten about the simply-connectedness of the faces. You have translated only the simply-connectedness of the polyhedron.

EPSILON: You are wrong. I demand that *all* the circuits should bound: even the o-circuits. I have translated 'simply-connectedness of a polyhedron' into 'all 1-circuits and 2-circuits bound'; and 'simply-connectedness of the faces' into 'all o-circuits bound'.

GAMMA: I do not follow you. What is a o-circuit?

EPSILON: A o-chain is any sum of vertices. A o-circuit any sum of vertices whose boundary is zero.

GAMMA: But what is the boundary of a vertex? There are no *minus* 1-dimensional polytopes!

EPSILON: Of course there are. Or, rather, there is one: the empty set.

GAMMA: You are mad!

ALPHA: He may not be mad. He is introducing a convention. I do not mind what conceptual tools he adopts. Let us see his results.

EPSILON: I do not use conventions, and my concepts are not 'tools'. The empty set *is* the *minus* 1-dimensional polytope. Its existence for me is certainly more obvious than the existence of, say, your dog.

TEACHER: No Platonic propaganda! Show how your 'bounding 0-circuits' translate 'simply-connected faces'.

EPSILON: If you once realise that the boundary of any vertex is the empty set, the rest is nothing. According to my earlier definition, the boundary of one vertex is the empty set, but the boundary of two vertices is zero, because of the *mod* 2 algebra. The boundary of three vertices is again the empty set, and so on. So even numbers of vertices are circuits, odd numbers of vertices are not.

GAMMA: So the point of your requirement that 0-circuits should bound amounts to the requirement that any two vertices must bound a 1-chain, or in ordinary language to the requirement that any two vertices must be connected by some system of edges. This of course rules out ring-shaped faces. This is indeed the requirement which we used to call the 'simply-connectedness of faces taken separately'.

EPSILON: You can scarcely deny that my language, which is the *natural* language reflecting the *essence* of polyhedra, shows for the first time the deeply rooted essential identity of formerly disconnected, isolated, *ad hoc* criteria!

GAMMA [*aside*]: What I can scarcely deny is that I am puzzled! That the way to this 'natural simplicity' should be littered with such complications really is rather strange.

ALPHA: Let me check that I understand. Do you say that all vertices have the same boundary: the empty set?

EPSILON: That is right.

ALPHA: And for you 'all vertices have the empty set' is an axiom, I assume; just as 'all faces have edges' or 'all edges have vertices'.

EPSILON: That is right.

ALPHA: But these axioms cannot possibly have an equal standing! The first is a convention, the last two are necessarily true!

TEACHER: The theorem has been translated. I want to see the proof.

EPSILON: Anon, Sir. Allow me a slight reformulation of the theorem to: '*All polyhedra in which circuits and bounding circuits coincide, are Eulerian*'.

TEACHER: Prove it.

EPSILON: Anon, Sir. I restate it.[1]

---

[1] 'Could you restate the problem? Could you restate it differently?' (Pólya [1945], inside cover).

BETA: But why? You have already translated all your terms which were a bit obscure into terms which are perfectly known!

EPSILON: That is true. But the translation I am about to produce is a very different one. I shall translate the set of my primitive terms into another set of primitive terms, which are still more basic.

BETA: So some of your perfectly known terms are better known than others!

TEACHER: Beta, do not constantly heckle Epsilon! Fix your attention on what he is doing and not on how he interprets what he is doing. Go on, Epsilon.

EPSILON: If we look more closely at my last formulation of the theorem we shall see that it is a theorem about the number of dimensions of certain vector spaces determined by the incidence matrix.

BETA: What?

EPSILON: Look at our concept of a chain, say a 1-chain. It is this:

$$x_1\theta_1 + x_2\theta_2 + \ldots + x_E\theta_E,$$

where $\theta_1, \ldots, \theta_E$ are the E edges, and $x_1, x_2, \ldots, x_E$ are either 0 or 1.

It is easy to see that the 1-chains form an E-dimensional vector-space over the field of residue-classes *modulo* 2. In general the k-chains form N_k-dimensional vector-spaces over the field of residue-classes *modulo* 2 (where N_k stands for the number of k-polytopes). The circuits form subspaces of the chain spaces and the bounding circuits again subspaces of the circuit spaces.

So my theorem in fact is that '*If the circuit-spaces and bounding circuit spaces coincide, the number of dimensions of the 0-chain space* minus *the number of dimensions of the 1-chain space* plus *the number of dimensions of the 2-chain space equals* 2'. This is the essence of Euler's theorem.

TEACHER: I like this reformulation which really showed the nature of your simple tools – just as you promised. You will now no doubt prove Euler's theorem by the simple methods of vector algebra. Let us see your proof.

2. Another Proof of the Conjecture

EPSILON: I decompose my theorem into two parts. The first states that the circuit spaces and bounding circuit spaces coincide if and only if the numbers of their dimensions coincide. The second states that if the circuit spaces and bounding circuit spaces have the same dimension, then the number of dimensions of the 0-chain space minus the number of dimensions of the 1-chain space *plus* the number of dimensions of the 2-chain space equals 2.

TEACHER: The first part is a trivially true theorem of vector algebra. Prove the second part.

EPSILON: Nothing is easier than that. I need only fall back on the definitions of the concepts involved.[1] First let us write out our incidence matrices. For instance let us take the incidence matrices of a tetrahedron *ABCD*, with edges *AD, BD, CD, BC, AC, AB* and faces *BCD, ACD, ABD, ABC*. The matrices are $\eta_{ij}^k = 1$ or 0, according as P_{k-1}^i does, or does not, belong to P_k^j. So our matrices are:

η^0	*A*	*B*	*C*	*D*
the empty set	1	1	1	1

η^1	*AD*	*BD*	*CD*	*BC*	*AC*	*AB*
A	1	0	0	0	1	1
B	0	1	0	1	0	1
C	0	0	1	1	1	0
D	1	1	1	0	0	0

η^2	*BCD*	*ACD*	*ABD*	*ABC*
AD	0	1	1	0
BD	1	0	1	0
CD	1	1	0	0
BC	1	0	0	1
AC	0	1	0	1
AB	0	0	1	1

η^3	*ABCD*
BCD	1
ACD	1
ABD	1
ABC	1

Now with the help of these matrices, the circuit spaces and the bounded circuit spaces can be easily characterised. We have already seen that the *k*-chains are really the vectors

$$\sum_{i=1}^{N_k} x_i P_i^k.$$

Now we defined the boundary of a P_j^k-polytope as

$$\sum_{i=1}^{N_{k-1}} \eta_{ij}^k P_i^{k-1}.$$

---

[1] 'To substitute mentally the definitions in place of the things defined' (Pascal [1659]).
'Go back to definitions' (Pólya [1945], inside cover and p. 84).

(This – like all the formulae which follow – is only a restatement of our old definition in symbolic notation.)

The boundary of a k-chain $\Sigma\, x_j P_j^k$ is

$$\sum_i \sum_j x_j \eta_{ij}^k P_i^{k-1}.$$

Now a k-chain $\Sigma\, x_j P_j^k$ is a k-circuit if and only if

(1) $\displaystyle\sum_{j=1}^{N_k} \eta_{ij}^k x_j = 0$ for each i.

A k-chain $\Sigma\, x_j P_j^k$ is a bounding k-circuit if and only if it is the boundary of some $(k+1)$-chain $\Sigma\, y_m P_m^{k+1}$, i.e. if and only if there exist coefficients $y_m (m = 1, \ldots, N_{k+1})$ such that

(2) $x_j = \Sigma\, y_m \eta_{jm}^{k+1}$.

Now it is obvious that the circuit space and the bounding circuit space are identical if and only if their dimensions are identical, i.e. if and only if the number of independent solutions of the N_{k-1} homogeneous linear equations (1) equals the number of independent solutions of the system of inhomogeneous linear equations (2). Now the first number is, according to the well known theorems of linear algebra, $N_k - \rho_k$ where ρ_k is the rank of $||\eta_{ij}^k||$; the second number is ρ_{k+1}.

So I have only to prove that if $N_k - \rho_k = \rho_{k+1}$ then $V - E + F = 2$.

LAMBDA: Or, 'If $N_k = \rho_k + \rho_{k+1}$ then $N_0 - N_1 + N_2 = 2$'. N_k are dimensions of certain vector spaces, ρ_k the ranks of certain matrices. This is no longer a theorem about polyhedra but about a certain set of multidimensional vector spaces.

EPSILON: I see you have just woken up. While you were asleep, I analysed our concepts of polyhedra and showed that they are *really* vector algebraic concepts. I translated the circle of ideas of the Euler-phenomenon into vector algebra, thus displaying their essence. Now I am certainly proving a theorem in vector algebra, which is a clear and distinct theory with perfectly known terms, neat and indubitable axioms, and with neat, indubitable proofs. For instance, look at the new trivial proof of our old much-discussed theorem: If $N_k = \rho_k + \rho_{k+1}$, then $N_0 - N_1 + N_2 = \rho_0 + \rho_1 - \rho_1 - \rho_2 + \rho_2 + \rho_3 = \rho_0 + \rho_3 = 1 + 1 = 2$. Who would dare to doubt the certainty of this theorem now? Thus I proved Euler's controversial theorem with indubitable certainty.[1]

ALPHA: But look here Epsilon, if we had accepted a rival convention

---

[1] This proof is due to Poincaré (see his [1899]).

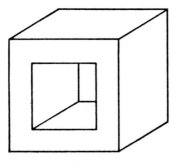

Fig. 29.

that the vertices have no boundary, the matrix η^0 for instance in the case of the tetrahedron would have been

$$\begin{array}{ccccc} \eta^0 & A & B & C & D \\ & 0 & 0 & 0 & 0 \end{array}$$

the rank ρ_0 would have been 0 and consequently $V - E + F = \rho_0 + \rho_3 = 1$. Do you not think your 'proof' relies too heavily on a convention? Did you not choose your convention only to save the theorem?

EPSILON: My axiom concerning ρ_0 was not a 'convention'. $\rho_0 = 1$ has in my language the very real meaning that each couple of vertices bounds, that is the network of edges is connected (ringshaped faces are thereby excluded). The expression 'convention' is utterly misleading. For polyhedra with simply connected faces, $\rho_0 = 1$ is *true*, $\rho_0 = 0$ is false.

ALPHA: Hmm. You seem to say that both $\rho_0 = 1$ and $\rho_0 = 0$ characterise some structure in vector-spaces. The difference is that $\rho_0 = 1$ has a real model in polyhedra with simply-connected faces, while the other has not.

3. Some Doubts about the Finality of the Proof. Translation Procedure and the Essentialist versus the Nominalist Approach to Definitions

TEACHER: Anyway, we have got the new proof. Is it final, though?

ALPHA: It is not. Take this polyhedron (fig. 29). It has two ringshaped faces, in the front and in the back, it can be pumped into a torus. And it has 16 vertices, 24 edges and 10 faces. Thus $V - E + F = 16 - 24 + 10 = 2$. It is Eulerian, but far from being simply-connected.

BETA: I do not think that this is an instance of the Descartes–Euler phenomenon. This is an instance of the Lhuilier phenomenon; that is: *for a polyhedron with k tunnels and m ringshaped faces $V - E + F = 2 - 2k + m$.*[1] For any polyhedron like this one with twice as many ringshaped faces as tunnels, $V - E + F = 2$, but that does not mean that it is Eulerian. And this Lhuilier-phenomenon explains at once why we could not easily get a necessary and sufficient condition – or mastertheorem – for the Descartes–Euler conjecture, because these Lhuilier instances intruded among the Eulerian ones.[2]

TEACHER: But Epsilon never promised finality, only more depth than we had achieved earlier. He has now fulfilled his promise to produce a proof which explains both the Eulerian character of ordinary polyhedra and the Eulerian character of star-polyhedra at one blow.

LAMBDA: This is true. He translated the requirement that the faces be simply-connected – that is that in the triangulating process each new diagonal should create a new face – in such a way that the idea of triangulation disappeared from it completely. In this new translation a face is simply-connected if all vertex-circuits bound in it – and *this* requirement holds for Eulerian star-polyhedra! And while we have difficulties in applying Jordan's intuitive (i.e. non-*star*-intuitive) concept of simply-connectedness of the polyhedron to star-polyhedra, in the Poincaré translation these difficulties disappear. Star-polyhedra, just like ordinary polyhedra, are sets of vertices, edges and faces plus an incidence-matrix; we are not concerned with the problem of a polyhedron's realization in a space which happens to be our material, three-dimensional, roughly Euclidean one. The small stellated dodecahedron for instance is not Eulerian: and it is not too difficult to trace 1-circuits on it which do not bound.

BETA: I find this interesting also from another aspect. Epsilon's proof is at once more rigorous and more embracing. Is there a necessary connection between these two?

EPSILON: I do not know. But while our Teacher claims only *more depth* for my proof, I am claiming *absolute certainty*.

KAPPA: Your theorem is as liable to be refuted by some imaginative concept-stretching as any previous conjecture.

EPSILON: You are wrong, Kappa, as I shall explain.[3]

ALPHA: Before you do let me raise a second question about your proof, or rather about the finality and certainty that you claim for it.

---

[1] See Lhuilier [1812–13a]. The relation was rediscovered about a dozen times between 1812 and 1890.
[2] See above, pp. 63 ff.   [3] See pp. 124–6.

Is the polyhedron in fact a model of your vector-algebraic structure? Are you sure that your translation of 'polyhedron' into vector theory was a *true* translation?

EPSILON: I have already said that it is true. If something startles you that is no reason for doubting it. 'I am following the great school of mathematicians who, in virtue of a series of startling definitions, have saved mathematics from the sceptics, and provided a rigid demonstration of its propositions.'[1]

TEACHER: I indeed think that this method of translation is the heart of the matter of the certainty and finality of Epsilon's proof. I think we should call it *translation-procedure*. But let us see, are there any other doubts?

GAMMA: Just one more. Say I accept that your deduction is infallible. Are you sure that you cannot deduce from your premises the negation of your theorem with the same infallibility?

EPSILON: All my premises are true. How can they possibly be inconsistent?

TEACHER: I appreciate your doubts. But I always prefer *one* counterexample to any number of doubts.

GAMMA: I wonder, does my cylinder not refute this new theorem?

EPSILON: Of course it does not. In the cylinder the empty set does not bound, and consequently $\rho_0 \neq 1$.

GAMMA: I see. You are right. This argument, put into your perfectly familiar, clear and distinct terms has convinced me at once.

EPSILON: I understand your sarcasm! Once before you queried my definitions. I then said they are, in fact, indubitably true axioms stating the essence of the concepts in question, with the help of infallibly clear and distinct intuition. I have thought about this since and I think I have to give up my Aristotelian view of definitions. When I define a vague term, I in fact replace it by a new one, and the old term serves only as an *abbreviation* of my new one.

ALPHA: Let me get this clear. What do you mean by 'definition': a replacement which is an operation from the left to the right or an abbreviation which is an operation from the right to the left?

---

[1] This is quoted from Ramsey [1931], p. 56. Only one word is changed, he says 'mathematical logicians' instead of 'mathematicians', but this is only because he did not understand that the procedure he described was not a novel characteristic of mathematical logic, but a feature of 'rigorous' mathematics from Cauchy on, and that the celebrated definitions of limit, continuity, and so on, proposed by Cauchy and improved by Weierstrass all fall in this line. I note that Russell also quotes this sentence from Ramsey (Russell [1959], p. 125).

EPSILON: I mean the abbreviation. I forget about the old meaning. I create freely the meaning of my terms while scrapping old vague terms. I also create my problems freely, while scrapping old obscure ones.

ALPHA: You cannot help being an extremist. But go on.

EPSILON: By this change in my programme I certainly gain one thing: one of your doubts is herewith eliminated. If definitions are abbreviations, then they cannot be false.

ALPHA: But you lose something which is much more important. You have to restrict your Euclidean programme to theories with perfectly known concepts, and when you want to pull theories with vague concepts into the scope of this programme, you cannot do this by your translational technique: as you said, you do not translate, rather you create new meaning. But even if you tried to *translate* the old meaning, some essential aspects of the original vague concept may get lost in this translation. The new clear concept may not serve for the solution of the problem for which the old concept was meant to serve.[1] If you regard your translation as infallible, or, if you consciously scrap the old meaning, both these extremes will yield the same result: you may push out the original problem into the limbo of the history of thought – which in fact you do not want to do.[2] So if you calm down, you have to admit that definition must have a touch of modified essentialism: it must preserve some relevant aspects of the old meaning, it must transfer relevant elements of meaning from left to right.[3]

---

[1] A classical example of a translation which did not satisfy the (usually implicit) adequacy criterion was the nineteenth-century definition of the area of a surface, which was knocked out by the Schwartz 'counterexample'.

The trouble is that adequacy criteria may change with the emergence of new problems which may occasion a change in the conceptual tool-cabinet. A paradigm case for such a change is the story of the concept of the integral. It is a shame of present mathematical education that students can quote exactly the different definitions of the Cauchy, Riemann, Lebesgue, etc. integrals, without knowing which problems they were invented to solve, or in the course of the solution of which problems they were discovered. As adequacy criteria change, definitions usually develop in such a way that the definition complying with all of the criteria becomes dominant. This could not happen to the definition of the integral, because of the inconsistency of the criteria – this is why the concept had to be split up. Proof-generated definitions play a decisive role even in building up translatory definitions in the Euclidean programme.

[2] This process is very characteristic of twentieth-century formalism.

[3] This trivial point is curiously enough missed by nominalists like Pascal and Popper. Pascal writes (*loc. cit.*): '...geometers and all those who operate methodically, impose names on things only to abridge discourse'. And Popper writes ([1945], volume 2, p. 14): 'In modern science only nominalist definitions occur, that is to say, shorthand symbols or labels are introduced to cut a long story short'. It is intriguing how nominalists and essentialists can each be blind to the rational kernel of the other's argument.

BETA: But even if Epsilon will accept this modified essentialism in definitions, the resignation from the essentialist approach will still be a huge withdrawal from his original Euclidean programme. Epsilon now says that there are Euclidean theories with perfectly known terms and infallible inferences – like arithmetic, geometry, logic, set theory I suppose, and he now makes the Euclidean programme consist of translating non-Euclidean theories with vague, obscure terms and uncertain inferences – like calculus and probability theory – into these already Euclidean theories, thus opening new avenues of development both of the underlying theories and of the originally non-Euclidean theories.

EPSILON: I shall call such an 'already Euclidean' or established theory a *dominant theory*.

GAMMA: I wonder what is the field of applicability of this shrunken programme? It certainly will not cover physics. You will never translate wave-mechanics into geometry. Epsilon wanted, 'in virtue of a series of startling definitions to save mathematics from the sceptics',[1] but what he saved was at best some crumbs.

BETA: I have a problem about those translatory definitions. They seem to appear as mere abbreviations in the dominant theory and thus there they are true 'by definition'. But they seem to be falsifiable if we regard them as referring to the non-Euclidean realm.[2]

EPSILON: That is right.

---

[1] See above, p. 121.

[2] The methodological importance of this difference has not yet been properly worked out. Pascal, the great advocate of abbreviatory definitions and the great opponent of the Aristotelian essentialist theory of definition, did not notice that to abandon essentialism is in fact to abandon the large-scale Euclidean programme. In the Euclidean programme one has to define all the terms that are 'only a bit obscure'. If this consists only of replacement of a vague term by an arbitrarily chosen precise one, one in fact abandons the original field of enquiry and turns to another. But Pascal certainly did not want this. Cauchy and Weierstrass were essentialists when carrying out the arithmetisation of mathematics; Russell was an essentialist when carrying out the logicisation of mathematics. All these men thought of their definitions of continuity, real numbers, integers and so on as capturing the essence of the concept involved. When stating the logical form of statements in ordinary language, i.e. translating ordinary language into artificial language, Russell thought – at least in his 'honeymoon period' ([1959], p. 73) – that he was guided by an infallible intuition. Popper, in his justified onslaught against essentialist definitions does not pay enough attention to the important problem of translatory definitions and I guess that this may account for what seems to me his unsatisfactory treatment of logical form in his [1947], p. 273. According to him (and here he follows Tarski) the definition of valid inference hinges *only* on the list of formative signs. *But validity of an intuitive inference depends also on the translation of the inference from ordinary (or arithmetical, geometrical, etc.) language into the logical language: it depends on the translation we adopt.*

BETA: It would be interesting to see how one falsifies such definitions.

THETA: I should like now to switch the discussion back to the question of the infallibility of Epsilon's deduction. Epsilon, do you still claim certainty for your theorem?

EPSILON: Certainly.

THETA: So you cannot imagine a counterexample to it?

EPSILON: As I told Kappa, my proof is infallible. There are no counterexamples to it.

THETA: Do you mean you would rule out counterexamples as monsters?

EPSILON: Not even a monster can refute it.

THETA: So you claim that whatever I substitute in the place of your perfectly known terms, the theorem remains true?

EPSILON: You can substitute anything in the place of the perfectly known terms which are *specific* to vector algebra.

THETA: I cannot replace your non-specific primitive terms, like 'all', 'and', '2' and so on?

EPSILON: No. But you can replace anything in the place of my *specific* perfectly known terms like 'vertex', 'edge', 'face' and so on. By this I think I clarified what I mean by refutation.

THETA: You did. But then you either can be refuted or you indeed did not do what you thought you did.

EPSILON: I do not understand your obscure hint.

THETA: You will, if you want to. Your characterisation of the idea of a counterexample seems reasonable. But if that is what a counterexample is, then the meaning of your 'perfectly well known terms' is immaterial. And this, if your claim is justified, is precisely the merit of your proof. A proof, if irrefutable, does not hinge – by the very concept of an irrefutable proof – on the meaning of the specific 'perfectly well known terms'. So the burden of your proof – if you are right – is fully borne by the meaning of the non-specific, underlying terms – in this case arithmetic, set-theory, logic – but not in the least by the meaning of your specific terms.

I shall call such proofs *formal proofs*, as they do not depend at all on the meaning of the specific terms. The *degree of formality* certainly depends on the non-specific terms. The perfectly known character of these terms – I shall call them formative terms – is very important indeed. By pinning down their meaning we state what can be accepted as counterexamples and what cannot. Thus we regulate the spate of counterexamples. If there are no counterexamples to the theorem, we shall call the theorem a *tautology*: in our case an arithmetico-set theoretical tautology.

ALPHA: We seem to have quite a gamut of tautologies according to our choice of quasilogical constants. But I see here a host of problems. First: how do we know of a tautology that it is a tautology?

KAPPA: *You will never know* beyond any possibility of doubt. But if you have *serious* doubts about a dominant theory then scrap it, and replace it by another dominant theory.[1]

---

* *Editors' note.* This section of the dialogue ends here in Lakatos's thesis. We should have tried to persuade Lakatos to continue the dialogue along the following lines:

THETA: But from what has just been said it seems to follow that if we can cast our proofs in systems in which the dominant theory is logic, then so long as we have no serious doubts about our logic, we shall be able to ensure the infallibility of our deductions and throw all the doubt not on the actual proof, but on the lemmas, on the antecedents of the theorem.

[1] Such changes in the dominant theory imply the reorganisation of all our knowledge. In antiquity the paradoxicality and, indeed, seeming inconsistency of arithmetic induced the Greeks to abandon arithmetic as the dominant theory and replace it by geometry. Their theory of proportions served the purpose of translating arithmetic into geometry. They were convinced that all astronomy, and all physics could be translated into geometry.

Descartes's great innovation was to replace geometry by algebra; maybe because he thought that in the dominant theory analysis itself should lead to truth.

The modern mathematical 'revolution of rigour' consisted in fact of the re-establishment of arithmetic as the dominant theory via the huge programme of the arithmetisation of mathematics which went on from Cauchy to Weierstrass. The theory of real numbers – felt to be artificial by quite a few working mathematicians – was the crucial step; analogous to the similarly 'artificial' theory of proportions of the Greeks.

Russell in turn made logic the dominant theory of all mathematics. The interpretation of the history of metamathematics as a search for a dominant theory may throw new light on the history of this subject, and one may be able to show that the Gödelian 'discovery' that the natural dominant theory for metamathematics is arithmetic, led straight to the present stage of inquiry, and opened new vistas both in arithmetic and in metamathematics.

Another example of a remarkable Euclidean translation was the modern embedding of probability theory into measure theory.

Dominant theories and the change of dominant theories also determine much of the development of science in general. The elaboration and then the breakdown of rational mechanics as the dominant theory of physics played a central role in modern history of science. The struggle of biology against being 'translated' into chemistry, the struggle of psychology against being translated into physiology, are intriguing features of the history of recent science. The translation procedures are vast reservoirs of problems, historical trends which represent huge patterns of thought at least as important as the Hegelian triad. Such translations usually speed up the development of both the dominant and the absorbed theory, but later the translation will become an impediment to further development as the weak spots of the translation come into the foreground.

EPSILON: I am glad that at least Theta finally caught on. My proof can in fact be cast in a system of which the dominant theory is logic. The conditional statement with all the lemmas incorporated as antecedents can be proved in this system, and we know that (relative to the given stock of formative 'logical' terms) there *are* no counterexamples to any statement which can be proved in this way. No matter how the descriptive terms are re-interpreted, this conditional statement will remain true.

LAMBDA: How do 'we know'?

EPSILON: We don't know for *certain* – it is an informal theorem about logic. But, moreover, we know that, presented with any alleged proof in such a system, we can check completely mechanically using a procedure which is guaranteed to produce an answer in a finite number of steps, whether or not it is indeed a proof. In such systems, then, your 'proof-analysis' reduces to a triviality.

ALPHA: But you would agree, Epsilon, that 'proof-analysis' retains its importance in informal mathematics; and that formal proofs are always translations of informal proofs and that the problems that have been raised about translation are very real.

LAMBDA: But anyway, Epsilon, how do we know that proof checking is always accurate?

EPSILON: Really Lambda, your unquenchable thirst for certainty is becoming tiresome! How many times do I have to tell you that we know nothing for certain? But your desire for certainty is making you raise very boring problems – and is blinding you to the interesting ones.

ANOTHER CASE-STUDY IN THE METHOD
OF PROOFS AND REFUTATIONS

1. Cauchy's Defence of the 'Principle of Continuity'

The method of proofs and refutations is a very general heuristic pattern of mathematical discovery. However, it seems that it was discovered only in the 1840s and even today seems paradoxical to many people; and certainly it is nowhere properly acknowledged. In this appendix I shall try to sketch the story of a proof-analysis in mathematical analysis and to trace the sources of resistance to the understanding and recognition of it. I first repeat the skeleton of the method of proofs and refutations, a method which I have already illustrated by my case-study of the Cauchy proof of the Descartes–Euler conjecture.

There is a simple pattern of mathematical discovery – or of the growth of informal mathematical theories. It consists of the following stages:[1]

(1) *Primitive conjecture.*

(2) *Proof (a rough thought-experiment or argument, decomposing the primitive conjecture into subconjectures or lemmas).*

(3) *'Global' counterexamples (counterexamples to the primitive conjecture) emerge.*

(4) *Proof re-examined: the 'guilty lemma' to which the global counterexample is a 'local' counterexample is spotted. This guilty lemma may have previously remained 'hidden' or may have been misidentified. Now it is made explicit, and built into the primitive conjecture as a condition. The theorem – the improved conjecture – supersedes the primitive conjecture with the new proof-generated concept as its paramount new feature.*[2]

---

[1] As I have stressed the actual historical pattern may deviate slightly from this heuristic pattern. Also the fourth stage may sometimes precede the third (even in the heuristic order) – an ingenious proof analysis may suggest the counterexample.

[2*] *Editors' note:* In other words this method consists (in part) of producing a series of statements $P_1, \ldots, P_n$ such that P_1 &... & P_n is supposed to be true of some domain of interesting objects and seems to imply the primitive conjecture C. This may turn out not to be the case – in other words we find cases in which C is false ('global counterexamples') but in which P_1 to P_n hold. This leads to the articulation of a new lemma

These four stages constitute the essential kernel of proof analysis. But there are some further standard stages which frequently occur:

(5) *Proofs of other theorems are examined to see if the newly found lemma or the new proof-generated concept occurs in them: this concept may be found lying at cross-roads of different proofs, and thus emerge as of basic importance.*

(6) *The hitherto accepted consequences of the original and now refuted conjecture are checked.*

(7) *Counterexamples are turned into new examples – new fields of inquiry open up.*

I should now like to consider another case-study. Here the *primitive conjecture* is that the limit of any convergent series of continuous functions is itself continuous. It was Cauchy who gave the first proof of this conjecture, whose truth had been taken for granted and assumed therefore not to be in need of any proof throughout the eighteenth century. It was regarded as the special case of the 'axiom' according to which 'what is true up to the limit is true at the limit'.[1] We find the conjecture and its proof in Cauchy's celebrated [1821] (p. 131).

Given that this 'conjecture' had hitherto been regarded as trivially true, why did Cauchy feel the need to prove it? Had someone criticized the conjecture?

As we shall see, the situation was not quite so simple. With the benefit of hindsight we can now see that counterexamples to the Cauchy conjecture had been provided by Fourier's work. Fourier's *Mémoire sur la Propagation de la Chaleur*[2] actually contains an example of what, according to present notions, is a convergent series of continuous functions which tends to a Cauchy discontinuous function, namely:

$$\cos x - \tfrac{1}{3} \cos 3x + \tfrac{1}{5} \cos 5x - \dots \qquad (1)$$

P_{n+1} which is also refuted by the counterexample ('local counterexample'). The original proof is thus replaced by a new one which can be summed up by the conditional statement

$$P_1 \And \dots \And P_n \And P_{n+1} \to C.$$

The (logical) truth of this conditional statement is no longer impugned by the counterexample (since the antecedent is now false in this case and hence the conditional statement true).

[1] Whewell [1858], 1. p. 152. Whewell is in 1858 at least ten years out of date. The principle stems from Leibniz's principle of continuity ([1687], p. 744). Boyer in his [1939], p. 256, quotes a characteristic restatement of the principle from Lhuilier [1786], p. 167.

[2] This *Mémoire* was awarded the *grand prix de mathématiques* for 1812, having been refereed by Laplace, Legendre and Lagrange. It was published only after Fourier's classical *Théorie de la Chaleur* which appeared in 1822, a year after Cauchy's textbook, but the content of the *Mémoire* was then already well known.

Fourier's own attitude to this series is, however, quite clear (and clearly different from this modern one):

(a) He states that it is everywhere convergent.

(b) He states that its limit function is composed of separate straight lines, each of which is parallel to the x-axis, and equal to the circumference [that is π]. These parallels are situated alternately above and below the axis, with a distance of $\pi/4$, and are joined by perpendiculars which themselves make part of the line.[1]

Fourier's words about the perpendiculars in the graph are telling. He considered these limit functions to be (in some sense) continuous. In fact, Fourier certainly regarded anything as a continuous function if its graph could be drawn with a pencil which is not lifted from the paper. Thus Fourier would not have regarded himself as having constructed counterexamples to Cauchy's continuity axiom.[2] It was only in the light of Cauchy's subsequent characterisation of continuity that the limit functions in some of Fourier's series came to be regarded as discontinuous, and thus that the series themselves came to be seen as counterexamples to Cauchy's conjecture. Given this new, and counterintuitive definition of continuity, Fourier's innocent continuous drawings seemed to become wicked counterexamples to the old, long established continuity principle.

Cauchy's definition certainly translated the homely concept of continuity into arithmetical language in such a way that 'ordinary

---

[1] Fourier, *op. cit.*, sections 177 and 178.

[2] After writing this I discovered that the term 'discontinuous' appears in roughly the Cauchy sense in some hitherto unpublished manuscripts of Poisson (1807) and of Fourier (1809), which were being studied by Dr. J. Ravetz, who kindly permitted me to look at his photostats. This certainly complicates my case, though it does not refute it. Fourier obviously had two different notions of continuity in mind at different times, and indeed these two different notions arise quite naturally from two different domains. If we interpret a function like:

$$\sin x - \tfrac{1}{2} \sin 2x + \tfrac{1}{3} \sin 3x - \ldots$$

as the initial position of a string, it will certainly be considered as continuous, and to cut out the perpendicular lines – as was to be required by Cauchy's definition – will seem unnatural. But if we interpret this function as, say, representing temperature along a wire, the function will seem obviously discontinuous. These considerations suggest two conjectures. Firstly, Cauchy's celebrated definition of continuity, which runs counter to the 'string-interpretation' of a function, may have been stimulated by Fourier's investigation of heat phenomena. Secondly, Fourier's insistence on the perpendiculars in the graphs of these (according to the 'heat-interpretation') discontinuous functions may have stemmed from an effort not to come into conflict with the Leibniz principle. *Editors' note*: For further information on Fourier's mathematics, see I. Grattan-Guinness (in collaboration with J. R. Ravetz), *Joseph Fourier, 1768–1830* (M.I.T. Press, 1972).

commonsense' could only be shocked.[1] What sort of continuity is it that implies that if we rotate the graph of a continuous function a little, it turns into a discontinuous one?[2]

So if we replace the intuitive concept of continuity by the Cauchy concept then (and only then!) does the axiom of continuity seem to be contradicted by Fourier's results. This looks like a strong, perhaps decisive, argument against Cauchy's new definitions (not only of continuity, but also other concepts like that of limit). No wonder then that Cauchy wanted to show that he could indeed prove the continuity axiom in his new interpretation of it, thereby providing the evidence that his definition satisfies this most stringent adequacy requirement. He succeeded in providing the proof – and thought he had thereby dealt a mortal blow to Fourier, that talented but woolly and unrigorous dilettante, who had unintentionally challenged his definition.

Of course if Cauchy's proof were correct, then Fourier's examples, despite appearances, could not be real *counter*examples. One way of showing that they were not real counterexamples would be to show that the series apparently converging to functions which were discontinuous in Cauchy's sense were not convergent at all!

And this was a plausible guess. Fourier himself was doubtful about the convergence of his series in these critical cases. He noticed that the convergence was slow: 'The convergence is not sufficiently rapid to produce an easy approximation, but it suffices for the truth of the equation.'[3]

With hindsight we can see that Cauchy's hope that in these critical cases Fourier's series do not converge (and thus do not represent the function) was also justified in a way by the following fact. Where the limit function is discontinuous, the series tends to $\frac{1}{2}[f(x+0)+f(x-0)]$, and not simply to $f(x)$. It tends to $f(x)$ only if $f(x) = \frac{1}{2}[f(x+0)+f(x-0)]$. But this was not known before 1829, and in fact general opinion was at

---

[1] That is string-commonsense or graph-commonsense.

[2●] *Editors' note:* What is violated here is, perhaps, not our intuitive notion of continuity, but rather our belief that any graph representing a function would still represent some function when slightly rotated. Fourier's curve is continuous from an intuitive point of view, and this intuition can still be accounted for by the ϵ, δ definition of continuity (with which Cauchy is usually credited); for Fourier's curve, complete with perpendiculars, is parametrically representable by two continuous functions.

[3] *Op. cit.*, section 177. This remark, of course, is a far cry from the discovery that the convergence is in these places infinitely slow, which was made only after 40 years experience in calculating Fourier series. And this discovery could not possibly be made before Dirichlet's decisive improvement on Fourier's conjecture showing that only those functions can be represented by Fourier series whose value at the discontinuities is $\frac{1}{2}[f(x+0)+f(x-0)]$.

first behind Fourier rather than Cauchy. Fourier's series seemed to work and when Abel, in 1826, five years after the publication of Cauchy's proof, mentioned in a footnote of his [1826b],[1] that there are 'exceptions' to Cauchy's theorem, this constituted a rather intriguing double victory: Fourier series were accepted, but so was Cauchy's startling definition of continuity and the theorem he had proved using it.

It was precisely in view of this double victory that it now seemed that there must be *exceptions* to the specific version of the principle of continuity we are considering, even though Cauchy had flawlessly proved it.

Cauchy must have reached the same conclusion as Abel for in the same year he gave, without of course giving up his characterisation of continuity, a proof of the convergence of the Fourier series.[2] He must have been very ill at ease with the situation however. The second volume of the *Cours d'Analyse* was never published. And, which is still more suspicious, he produced no further editions of the first volume, allowing his pupil Moigno, when the pressure for a textbook had become too great, to publish his notes of his lectures.[3]

Given that Fourier's examples were now interpreted as counter-examples, the puzzle was evident: how could a proved theorem be false, or 'suffer exceptions'? We have already discussed how people in the same period were puzzled by the 'exceptions' to the Euler theorem despite the fact that it had been proved.

2. Seidel's Proof and the Proof-Generated Concept of Uniform Convergence

Everybody felt that this Cauchy-Fourier case was not just a harmless puzzle, but a fatal blemish on the whole of the new 'rigorous' mathematics. Dirichlet in his celebrated papers about Fourier series,[4] while preoccupied with showing exactly *how* convergent series of continuous functions represent discontinuous functions, and while obviously very much aware of the Cauchy version of the continuity principle, did not mention the obvious contradiction at all.

It was left to Seidel at last to solve the riddle by spotting the guilty hidden lemma in Cauchy's proof.[5] But this happened only in 1847. Why did it take so long? To answer this question we shall have to look at Seidel's celebrated discovery a little more closely.

---

[1] Abel [1826b], p. 316.
[2] Cauchy [1826]. The proof is based on an incorrigibly false assumption (see e.g. Riemann, [1868]).   [3] Moigno [1840–1].   [4] Dirichlet [1829].   [5] Seidel [1847].

Let $\Sigma f_n(x)$ be a convergent series of continuous functions and, for any n, define $S_n(x) = \sum\limits_{m=0}^{n} f_m(x)$ and $r_n(x) = \sum\limits_{m=n+1}^{\infty} f_n(x)$. Then the gist of Cauchy's proof is the inference from the premise:

Given any $\epsilon > 0$:

(1) there is δ such that for any b, if $|b| < \delta$, then $|S_n(x+b) - S_n(x)| < \epsilon$ (there is such a δ because of the continuity of $S_n(x)$);

(2) there is an N, such that $|r_n(x)| < \epsilon$ for all $n \geqslant N$ (there is such an N because of the convergence of $\Sigma f_n(x)$);

(3) there is an N' such that $|r_n(x+b)| < \epsilon$ for all $n \geqslant N'$ (there is such an N' because of the convergence of $\Sigma f_n(x+b)$);

to the conclusion that:

$$
\begin{aligned}
|f(x+b) - f(x)| &= |S_n(x+b) + r_n(x+b) - S_n(x) - r_n(x)| \\
&\leqslant |S_n(x+b) - S_n(x)| + |r_n(x)| + |r_n(x+b)| \\
&< 3\epsilon, \text{ for all } b < \delta
\end{aligned}
$$

Now the global counterexamples provided by series of continuous functions which converge to Cauchy-discontinuous functions show that something is wrong with this (roughly stated) argument. But where is the guilty lemma?

A slightly more careful proof analysis (using the same symbols as before, but making explicit the functional dependencies of some of the quantities) produces the following inference:

(1') $|S_n(x+b) - S_n(x)| < \epsilon$ if $b < \delta(\epsilon, x, n)$

(2') $|r_n(x)| < \epsilon$ if $n > N(\epsilon, x)$

(3') $|r_n(x+b)| < \epsilon$ if $n > N(\epsilon, x+b)$

therefore

$$
|S_n(x+b) + r_n(x+b) - S_n(x) - r_n(x)| = |f(x+b) - f(x)| < 3\epsilon
$$

if $n > \max_z N(\epsilon, z)$ and $b < \delta(\epsilon, x, n)$.

The hidden lemma is that this maximum, $\max_z N(\epsilon, z)$, should exist for any fixed ϵ. This is what came to be called the requirement of *uniform convergence*.

There were probably three major impediments in the way of making this discovery.

The *first* was Cauchy's loose usage of 'infinitely small' quantities.[1] The *second* was that even if some mathematicians had noticed that the

[1] This prevented Cauchy from giving a clear critical appraisal of his old proof and even from formulating his theorem clearly in his [1853] (pp. 454-9).

assumption of the existence of a maximum of an infinite set of Ns is involved in this proof, they may very well have made it without a second thought. Existence proofs in maximum problems occur first in the Weierstrass school. But the *third* and main obstacle was the prevalence of Euclidean methodology – this good and evil spirit of early nineteenth century mathematics.

But before discussing this in general let us see how Abel solves the problem posed for the Cauchy theorem by the Fourier counter-examples. I shall show that he solves it (or rather 'solves' it) by the primitive 'exception-barring' method.[1]

3. Abel's Exception-Barring Method

Abel states the problem, which I claim to be the basic background problem of his celebrated paper on the binomial series,[2] only in a footnote. He writes: 'It seems to me that there are some exceptions to Cauchy's theorem', and immediately gives the example of the series

$$\sin \phi - \tfrac{1}{2} \sin 2\phi + \tfrac{1}{3} \sin 3\phi - \ldots [3]$$

Abel adds that 'as it is known, there are many more examples like this'. His response to these counterexamples is to start guessing: 'What is the safe domain of Cauchy's theorem?'

His answer to this question is this: the domain of validity of the theorems of analysis in general, and that of the theorems about the continuity of the limit function in particular, is restricted to power series. All the known exceptions to this basic continuity principle were trigonometrical series, and so he proposed to withdraw analysis to within the safe boundaries of power series, thus leaving behind Fourier's cherished trigonometrical series as an uncontrollable jungle – where exceptions are the norm and successes miracles.

In a letter to Hansteen dated 29 March 1826, Abel characterised 'miserable Eulerian induction' as a method which leads to false and unfounded generalisations and he asks what the reason is for such procedures having in fact led to so *few* calamities. His answer is

To my mind the reason is that in analysis one is largely concerned with functions that can be represented by power-series. As soon as other functions enter – and this happens but rarely – then [induction] does not work any more and an

---

[1] See above, pp. 24–30.   [2] Abel [1826b], p. 316.
[3] Abel fails to mention that precisely this example had already been mentioned in this context by Fourier.

infinite number of incorrect theorems arise from these false conclusions, one leading to the others. I have investigated several of these and I was lucky enough to solve the problem...[1]

In Abel's paper, we find his famous theorem – which, I claim, stemmed from his grappling with the classical metaphysical principle of Leibniz – in the following restricted form:

If the series

$$f\alpha = v_0 + v_1\alpha + v_2\alpha^2 + \ldots + v_m\alpha^m + \ldots$$

is convergent for a given value δ of α, it will also converge for every value smaller than δ, and for steadily decreasing values of β, the function $f(\alpha - \beta)$ will approach the limit $f\alpha$ indefinitely, provided that α is smaller than or equal to δ.[2]

Modern rationalist historians of mathematics who consider the history of mathematics as the history of a homogeneous growth of knowledge on the basis of unchanging methodology, assume that anyone who discovers a global counterexample and proposes a new conjecture which is not subject to refutation by the counterexample in question, has automatically discovered the corresponding hidden lemma and proof-generated concept. In this way such students of history attribute the discovery of uniform convergence to Abel. So in the authoritative *Encyclopädie der Mathematischen Wissenschaften*, Pringsheim says that Abel 'demonstrated the existence of the property today called uniform convergence'.[3] Hardy shares Pringsheim's view. In his [1918] paper he says that 'the idea of uniform convergence is

---

[1] Letter to Hansteen ([1826a]). The rest of the letter is also interesting and reflects Abel's exception-barring method: 'When one proceeds by a general method, it is not too difficult; but I have had to be very circumspect, for propositions once accepted without rigorous proof (i.e. without any proof) are so rooted within me that I at each moment risk using them without further examination.' Thus Abel checked these general conjectures one after the other and tried to guess the domain of their validity.

This Cartesian self-imposed restriction to the absolutely clear power-series explains Abel's particular concern about the rigorous treatment of the Taylor-expansion: 'Taylor's theorem, the basis of all the infinitesimal calculus is not better founded. I have only found one rigorous demonstration and that is M. Cauchy's in his *Résumé des leçons sur le calcul infinitesimal*, where he demonstrated that one will have

$$\phi(x+a) = \phi(x) + a\phi'(x) + a^2\phi''(x) + \ldots$$

as long as the series is convergent; but one employs it without attention in all cases.' (Letter to Holmboë [1825].)

[2] Abel [1826b], I. p. 314. The text is a retranslation from German, (Crelle translated the original French into German).* *Editors' note:* It seems that Abel forgot the modulus sign around α.   [3] Pringsheim [1916], p. 34.

present implicitly in Abel's proof of his celebrated theorem'.[1] Bourbaki is even more explicitly false: according to him, Cauchy

> did not at first perceive the distinction between simple convergence and uniform convergence, and considered himself able to demonstrate that every convergent series of continuous functions has as its sum a continuous function. The error was almost as soon revealed by Abel, who proved at the same time that every complete [?] series is continuous in the interior of its interval of convergence by the reasoning which has become classical and which uses essentially, in this particular, the idea of uniform convergence. It only remained to disentangle the latter in a general manner, which was done independently by Stokes and Seidel in 1847-8 and by Cauchy himself in 1853.[2]

So many sentences, so many mistakes. Abel did not reveal Cauchy's mistake in identifying the two sorts of convergences. His proof does not exploit the concept of uniform convergence any more than does Cauchy's. Abel's and Seidel's results are not in the relation of 'special' and 'general' – they are on quite different levels. Abel did not even notice that it is not the domain of eligible functions which has to be restricted, but rather the way they converge! *In fact for Abel there is only one sort of convergence, the simple one*; and the secret of the sham certainty of his proof lies in his cautious (and lucky) *zero-definitions*:[3] as we now know, in the case of power series, simple convergence coincides with uniform convergence![4]

---

[1] Hardy [1918], p. 148.

[2] Bourbaki [1949], p. 65 and [1960], p. 228.   [3] Cf. above, pp. 24-30.

[4] There were two mathematicians who noticed that Abel's proof was not quite flawless. One was Abel himself, who comes to grips with the problem again – without success – in his posthumously published paper 'Sur les Séries' ([1881], p. 202). The other was Sylow, the coeditor of the second edition of Abel's Collected Works. He added a critical footnote to the theorem, in which he pointed out that we have to require uniform convergence in the proof and not simple convergence, as Abel does. But he did not use the term 'uniform convergence' about which he did not seem to know, (the second edition of Jordan's *Cours d'Analyse* had not then appeared) and he referred instead to a later generalisation of du Bois-Reymond, which only shows that even he did not see clearly the nature of the flaw. Reiff, in his [1889], rejected Sylow's criticism with the naive argument that Abel's theorem is valid. Reiff says that while Cauchy was the founder of the theory of convergence, Abel was the founder of the theory of the continuity of series:

> Briefly summarizing the achievement of Cauchy and of Abel, we can say: Cauchy discovered the theory of the convergence and divergence of infinite series in his *Analyse Algébrique*, and Abel discovered the theory of the continuity of series in his *Treatise on the Binomial Series*. ([1889], pp. 178-9.)

To say this in 1889 was certainly a piece of pompous ignorance.

But of course the validity of Abel's theorem is due to the very narrow zero-definition, and not to the proof. Abel's paper was later published in *Ostwald's Klassiker* (Nr. 71), Leipzig, 1895. In the notes Sylow's remarks are reproduced without any comment.

Whilst I am criticizing the historians I should just mention that the first counterexample to Cauchy's theorem has generally been attributed to Abel. That it occurs in Fourier was noticed only by Jourdain. But he, in the ahistorical spirit already noted, draws from this fact the consequence that Fourier, for whom Jourdain had a great admiration, came close to discovering the concept of uniform convergence.[1] The point that a counterexample may have to fight for recognition, and when recognised it still may not lead automatically to the hidden lemma and thereby to the proof-generated concept in question, has been missed by all historians so far.

4. Obstacles in the Way of the Discovery of the Method of Proof-Analysis

But now let us return to the main problem. Why did the leading mathematicians from 1821 to 1847 fail to find the simple flaw in Cauchy's proof and improve both the proof-analysis and the theorem?

The first reply is that they did not know about the method of proofs and refutations. They did not know that after the discovery of a counterexample they had to analyse their proof carefully and try to find the guilty lemma. They dealt with global counterexamples with the help of the heuristically sterile exception-barring method.

In fact, Seidel discovered the proof-generated concept of uniform convergence and the method of proofs and refutations at one blow. He was fully conscious of his methodological discovery[2] which he stated in his paper with great clarity:

Starting from the certainty just achieved, that the theorem is not universally valid, and hence that its proof must rest on some extra hidden assumption, one then subjects the proof to a more detailed analysis. It is not very difficult to discover the hidden hypothesis. One can then infer backwards that this condition expressed by the hypothesis is not satisfied by series which represent discontinuous functions, since only thus can the agreement between the otherwise correct proof sequence, and what has been on the other hand established, be restored.[3]

What prevented the generation before Seidel from discovering this? The main reason (which we already mentioned) was the prevalence of Euclidean methodology.

---

[1] Jourdain [1912], 2, p. 527.

[2] Rationalists doubt that there are methodological discoveries at all. They think that method is unchanging, eternal. Indeed methodological discoverers are very badly treated. Before their method is accepted it is treated like a cranky theory; after, it is treated as a trivial commonplace.　　　　　[3] Seidel [1847], p. 383.

The Cauchy revolution of rigour was motivated by a conscious attempt to apply Euclidean methodology to the Calculus.[1] He and his followers thought that this was how they could introduce light to dispel the 'tremendous obscurity of analysis'.[2] Cauchy proceeded in the spirit of Pascal's rules: he first set out to define the obscure terms of analysis – like limit, convergence, continuity, etc. – in the perfectly familiar terms of arithmetic, and then he went on to prove everything that had not previously been proved, or that was not perfectly obvious. Now in the Euclidean framework there is no point trying to prove what is false, so Cauchy had first to improve the extant body of mathematical conjectures by jettisoning the false rubbish. In order to improve the conjectures, he applied the method of looking out for exceptions and restricting the domain of validity of the original, rashly stated conjectures to a safe field, i.e. he applied the exception-barring method.[3]

A writer in the 1865 edition of the *Larousse* (probably Catalan) rather sarcastically characterised Cauchy's search for counterexamples. He wrote:

He has introduced into science only negative doctrines...it is in fact almost always the negative aspect of the truth which he came to discover, that he takes care to make evident: if he had found some gold in whiting, he would have announced to the world that chalk is not *exclusively* formed of carbonate of lime.

A part of a letter which Abel wrote to Holmboë is further evidence of this new heartsearching mood of the Cauchy school:

I have begun to examine the most important rules which (at present) we ordinarily sanction in this respect, and to show in which cases they are not proper. This goes well enough and interests me infinitely.[4]

What was considered by the rigourists to be hopeless rubbish, such as conjectures about sums of divergent series, was duly committed to the flames.[5] 'Divergent series are', wrote Abel, 'the work of the devil'. They only cause 'calamities and paradoxicalities'.[6]

But while constantly endeavouring to improve their conjectures by

---

[1] 'As for methods, I have had to give them all the rigour that one demands in geometry, so as never to resort to reasons drawn from the generality of algebra.' (Cauchy [1821], Introduction.)   [2] Abel [1826a], p. 263.

[3] 'To bring useful restrictions to too extended assertions.' (Cauchy, [1821].)

[4] Abel [1825], p. 258.

[5] Contemporaries certainly regarded this purge as 'a little harsh'. (Cauchy, [1821], Introduction.)   [6] Abel [1825], p. 257.

exception-barring, the idea of *improving* by proving never occurred to them. The two activities of guessing and proving are rigidly separated in the Euclidean tradition. The idea of a proof which deserves its name and still is not conclusive was alien to the rigourists. Counter-examples were regarded as grave and disastrous blemishes: they showed that a conjecture was wrong and that one had to start proving again from scratch.

This was understandable in view of the fact that in the eighteenth century pieces of shabby inductive reasoning were called proofs.[1] But there was no way of improving *these* 'proofs'. They were rightly scrapped as 'not rigorous proofs – that means, no proofs at all'.[2] *Inductive argument was fallible – therefore it was committed to the flames. Deductive argument took its place – because it was held to be infallible.* 'I make all uncertainty disappear', announced Cauchy.[3] It is against this background that the refutation of Cauchy's 'rigorously' proved theorem has to be appreciated. And this refutation was not an isolated case. Cauchy's rigorous proof of the Euler formula was, as we have seen, followed likewise by papers stating the well known 'exceptions'.

There were only two ways out: either to revise the whole infallibilist philosophy of mathematics underlying the Euclidean method, or somehow to hush up the problem. Let us first see what would be involved in revising the infallibilist approach. One would certainly have to give up the idea that all mathematics can be reduced to in-dubitably true trivialities, that there are statements about which our truth-intuition cannot possibly be mistaken. One had to give up the idea that our deductive, inferential intuition is infallible. Only these two admissions could open the way to the free development of the method of proofs and refutations and its application to the critical appraisal of deductive argument and to the problem of dealing with counterexamples.[4]

---

[1] The eighteenth-century 'formalism' was sheer inductivism. Cf. p. 133, Cauchy rejects in the Preface of his [1821] inductions which are only 'appropriate to sometimes present the truth'.

[2] Abel, [1826a], p. 263. For Cauchy and Abel 'rigorous' means deductive, as opposed to inductive. [3] Cauchy [1821], Introduction.

[4]* *Editors' note:* This passage seems to us mistaken and we have no doubt that Lakatos, who came to have the highest regard for formal deductive logic, would himself have changed it. First order logic has arrived at a characterisation of the validity of an inference which (relative to a characterisation of the 'logical' terms of a language) does make valid inference essentially infallible. Thus, one need make only the first of the two admissions mentioned by Lakatos. By a sufficiently good 'proof analysis' all the doubt can be thrown onto the *axioms* (or antecedents of the theorem) leaving none on the *proof* itself. The method of proofs and refutations is by no means invalidated (as is suggested

As long as a counterexample was a blemish not only to a theorem but to the mathematician who advocated it, as long as there were only proofs or non-proofs, but no sound proofs with weak spots, mathematical criticism was barred. It was the infallibilist philosophical background of Euclidean method that bred the authoritarian traditional patterns in mathematics, that prevented publication and discussion of conjectures, that made impossible the rise of mathematical criticism. Literary criticism can exist because we can appreciate a poem without considering it to be perfect; mathematical or scientific criticism cannot exist while we only appreciate a mathematical or scientific result if it yields perfect truth. A proof is a proof only if it proves; and it either proves or it does not. The idea – expressed so clearly by Seidel – that a proof can be respectable without being flawless, was a revolutionary one in 1847, and, unfortunately, still sounds revolutionary today.

It is no coincidence that the discovery of the method of proofs and refutations occurred in the 1840s, when the breakdown of Newtonian optics (through the work of Fresnel in the 1810s and 1820s), and the discovery of non-Euclidean geometries (by Lobatschewsky in 1829 and Bolyai in 1832) shattered infallibilist conceit.[1]

in the text) by refusing to make the second of these admissions: indeed it may be by this method that proofs are improved so that all the assumptions that have to be made in order that the proof be valid, are made explicit.

[1] In the same decade Hegel's philosophy offered both a radical break with its infallibilist predecessors and a powerful start for a thoroughly novel approach to knowledge. (Hegel and Popper represent the only fallibilist traditions in modern philosophy, but even they both made the mistake of reserving a privileged infallible status for mathematics.) A passage from de Morgan shows the new fallibilist mood of the forties:

'A disposition sometimes appears to reject all that offers any difficulty, or does not give all its conclusions without any trouble in examination of apparent contradictions. If by this it be meant that nothing should be permanently used, and implicitly trusted, which is not true to the full extent of the assertion made, I, for one, should offer no opposition to so rational a course. But if it be implied that nothing should be produced to the student, with or without warning, which cannot be understood in all its generality, I should, with deference, protest against a restriction which would tend, in my opinion, not only to give false views of what is actually known, but to stop the progress of discovery. It is not true, out of geometry, that the mathematical sciences are, *in all their parts*, those models of finished accuracy which many suppose. The extreme boundaries of *analysis* have always been as imperfectly understood as the tract beyond the boundaries was absolutely unknown. But the way to enlarge the settled country has not been by keeping within it, [this remark is against the exception-barring method] but by making voyages of discovery, and I am perfectly convinced that the *student* should be exercised in this manner; that is, that he should be taught how to examine the boundary, as well as how to cultivate the interior. I have therefore never scrupled, in the latter part of the work, to use methods which I will not call doubtful, because they are presented as unfinished, and because the doubt is that of an expectant learner, not of an unsatisfied critic. Experience has often shown that the defective conclusion has been rendered

Before the discovery of the method of proofs and refutations the problem posed by the succession of counterexamples to a 'rigorously proved' theorem could be 'solved' only by the exception-barring method. *The proof proves the theorem, but it leaves the question open of what is the theorem's domain of validity. We can determine this domain by stating and carefully excluding the 'exceptions' (this euphemism is characteristic of the period). These exceptions are then written into the formulation of the theorem.*

The dominance of the exception-barring method shows how the Euclidean method can, in certain crucial problem situations, have deleterious effects on the development of mathematics. Most of these problem situations occur in growing mathematical theories, where growing concepts are the vehicles of progress, where the most exciting developments come from exploring the boundary regions of concepts, from stretching them, and from differentiating formerly undifferentiated concepts. In these growing theories intuition is inexperienced, it stumbles and errs. There is no theory which has not passed through such a period of growth; moreover, this period is the most exciting from the historical point of view and should be the most important from the teaching point of view. These periods cannot be properly understood without understanding the method of proofs and refutations, without adopting a fallibilist approach.

This is why Euclid has been the evil genius particularly for the history of mathematics and for the teaching of mathematics, both on the introductory and the creative levels.[1]

intelligible and rigorous by persevering thought, but who can give it to conclusions which are never allowed to come before him? The effect of exclusive attention to those parts of mathematics which offer no scope for the discussion of doubtful points is a distaste for modes of proceedings which are absolutely necessary to the extension of analysis. If the cultivation of the higher parts of mathematics were left to persons trained for the purpose, there might be some show of reason for keeping out of the ordinary student's reach, not only the unsettled, but even the purely speculative parts of the abstract sciences; reserving them for those persons whose business it would then be to render the former clear and the latter applicable. As it is, however, the few in this country who pay attention to any difficulty of mathematics for its own sake come to their pursuit through the casualties of taste or circumstances; and the number of such casualties should be increased by allowing all students whose capacity will let them read on the higher branches of applied mathematics, to have each his chance of being led to the cultivation of those parts of analysis on which rather depends its future progress than its present use in the sciences of matter.' (de Morgan [1842], p. vii).

[1] According to R. B. Braithwaite, 'the good genius of mathematics and of unselfconscious science, Euclid has been the evil genius of philosophy of science – and indeed of metaphysics'. (Braithwaite [1953], p. 353.) This statement, however, originates in a static logicist conception of mathematics.

Note: In this appendix the supplementary stages 5, 6, and 7 (cf. p. 128) of the method of proofs and refutations have not been discussed. I would just mention here that a methodical hunt for uniform convergence in other proofs (stage 5) would very quickly have yielded the refutation and improvement of another theorem proved by Cauchy: the theorem that the integral of the limit of any convergent series of continuous functions is the limit of the sequence of the integrals of the terms, or briefly, that in the case of series of continuous functions, the limit and the integral-operations can be interchanged. This had been uncontested throughout the eighteenth century, and even Gauss applied it without giving it a second thought. (See Gauss [1813], Knopp [1928] and Bell [1945].)

Now it did not occur to Seidel, who discovered uniform convergence in 1847, to look at other proofs to see if it had been implicitly assumed there. Stokes, who discovered uniform convergence in the same year – though not with the help of the method of proofs and refutations – uses in this same paper the false theorem about integration of series, referring to Moigno (Stokes [1848]). (Stokes made another mistake: he thought he had proved that uniform convergence was not only sufficient but necessary for the continuity of the limit function.)

This delay in discovering that the proof that the integration of series also depends on the assumption of uniform convergence may have been due to the fact that this primitive conjecture was refuted by a concrete counterexample only in 1875 (Darboux [1875]), by which date proof-analysis had already traced uniform convergence in the proof without the analysis being catalysed by a counterexample. The hunt for uniform convergence once fully under way with Weierstrass at its head soon discovered the concept in proofs concerning term by term differentiation, double limits, etc.

The *sixth stage* is to check the hitherto accepted consequences of the refuted primitive conjecture. Can we rescue these consequences, or does the refutation of the lemma lead to a disastrous holocaust? Term by term integration, for instance, was a cornerstone of the Dirichlet proof of Fourier's conjecture. Du Bois-Reymond describes the situation in dramatic terms: the theory of trigonometric series is 'cut to the heart', its two key theorems 'have had the ground cut from under them' and

with one blow the general theory was pushed back to the state in which it had been before Dirichlet, back even before Fourier.

(du Bois-Reymond [1875], p. 120.) It makes an intriguing study to see how the 'lost ground' has been regained.

In this process a spate of counterexamples was unearthed. But their study – the *seventh stage* of the method – started only in the last years of the century. (E.g. Young's work on the classification and distribution of points of non-uniform convergence; Young [1903–4].)

THE DEDUCTIVIST VERSUS THE
HEURISTIC APPROACH

1. The Deductivist Approach

Euclidean methodology has developed a certain obligatory style of presentation. I shall refer to this as 'deductivist style'. This style starts with a painstakingly stated list of *axioms*, *lemmas* and/or *definitions*. The axioms and definitions frequently look artificial and mystifyingly complicated. One is never told how these complications arose. The list of axioms and definitions is followed by the carefully worded *theorems*. These are loaded with heavy-going conditions; it seems impossible that anyone should ever have guessed them. The theorem is followed by the *proof*.

The student of mathematics is obliged, according to the Euclidean ritual, to attend this conjuring act without asking questions either about the background or about how this sleight-of-hand is performed. If the student by chance discovers that some of the unseemly definitions are proof-generated, if he simply wonders how these definitions, lemmas and the theorem can possibly precede the proof, the conjuror will ostracize him for this display of mathematical immaturity.[1]

In deductivist style, all propositions are true and all inferences valid. Mathematics is presented as an ever-increasing set of eternal, immutable truths. Counterexamples, refutations, criticism cannot possibly enter. An authoritarian air is secured for the subject by beginning with disguised monster-barring and proof-generated definitions and with the fully-fledged theorem, and by suppressing the primitive conjecture, the refutations, and the criticism of the proof. Deductivist style hides the struggle, hides the adventure. The whole story vanishes, the successive tentative formulations of the theorem in the course of the proof-procedure are doomed to oblivion while the end result is exalted into sacred infallibility.[2]

---

[1] Some textbooks claim that they do not expect the reader to have any previous knowledge, only a certain mathematical maturity. This frequently means that they expect the reader to be endowed by nature with the 'ability' to take a Euclidean argument without any unnatural interest in the problem-background, in the heuristic behind the argument.

[2] It has not yet been sufficiently realised that present mathematical and scientific education

Some who defend deductivist style claim that deduction is *the* heuristic pattern in mathematics, that the logic of discovery is deduction.[1] Others realise that this is not true, but draw from this realisation the conclusion that mathematical discovery is a completely non-rational affair. Thus they will claim that although mathematical discovery does not proceed deductively, if we want our presentation of mathematical discoveries to proceed rationally, it must proceed in the deductivist style.[2]

is a hotbed of authoritarianism and is the worst enemy of independent and critical thought. While in mathematics this authoritarianism follows the *deductivist* pattern just described, in science it operates through the *inductivist* pattern.

There is a longstanding tradition of inductivist style in science. An ideal paper written in this style starts with the painstaking description of the layout of the experiment, followed by the description of the experiment and its result. A 'generalisation' may conclude the paper. The problem-situation, the conjecture which the experiment had to test, is hidden away. The author boasts of an empty, virgin mind. The paper will be understood only by the few who actually know the problem-situation. – Inductivist style reflects the pretence that the scientist starts his investigation with an empty mind whereas in fact he starts with a mind full of ideas. This game can only be played – not always with success – by and for a selected guild of experts. Inductivist style, just like its deductivist twin (not counterpart!), while claiming objectivity, in fact fosters a private guild-language, atomises science, suffocates criticism, makes science authoritarian. Counterexamples can never occur in such presentation: one starts with observations (not a theory), and obviously unless one has a prior theory one cannot observe counterexamples.

[1] These people claim that mathematicians start with an empty mind, set up their axioms and definitions at their pleasure, in the course of a playful free creative activity, and it is only at a later stage that they deduce the theorems from these axioms and definitions. If in some interpretation the axioms are true, the theorems will all be true. The mathematical conveyor-belt of truth cannot fail. After our case-study in the proof-procedure this can be ruled out as an argument for the defence of the deductivist style in general – if we do not accept the restriction of mathematics to formal systems.

Now while Popper showed that those who claim that induction is the logic of scientific discovery are wrong, these essays intend to show that those who claim that deduction is the logic of mathematical discovery are wrong. While Popper criticised inductivist style, these essays try to criticise deductivist style.

[2] This doctrine is an essential part of most brands of formalist philosophies of mathematics. Formalists, when talking about discovery, discriminate the *context of discovery* and the *context of justification*. 'The context of discovery is left to psychological analysis, whereas logic is concerned with the context of justification.' (Reichenbach [1947], p. 2.) A similar view can be found in R. B. Braithwaite's [1953], p. 27, and even in K. R. Popper's [1959], pp. 31–2, and in his [1935]. Popper, when (in fact in 1934) dividing the aspects of discovery between psychology and logic in such a way that no place was left for heuristic as an independent field of inquiry, obviously had not then realised that his 'logic of discovery' was more than just the *strictly logical* pattern of the progress of science. This is the source of the paradoxicality of the title of his book, the thesis of which seems to be double-faced: (*a*) there is no logic of scientific discovery – Bacon and Descartes were both mistaken; (*b*) the logic of scientific discovery is the logic of conjectures and refutations. The solution of this paradox is at hand: (*a*) there is no *infallibilist* logic of scientific discovery, one which would infallibly lead to results;

So we have nowadays two arguments for deductivist style. One is based on the idea that heuristic is rational and deductivist. The second argument is based on the idea that heuristic is not deductivist, but also not rational.

There is also a third argument. Some working mathematicians who do not like logicians, philosophers and other cranks interfering in their work, usually say that the introduction of heuristic style would require the rewriting of textbooks, and would make them so long that one could never read them to the end. Papers would become much longer too.[1] The answer to this pedestrian argument is: let us try.

2. The Heuristic Approach. Proof-Generated Concepts

This section will contain brief heuristic analyses of some mathematically important proof-generated concepts. It is hoped these analyses will show the advantage of introducing heuristic elements into mathematical style.

As has already been mentioned, deductivist style tears the proof-generated definitions off their 'proof-ancestors', presents them out of the blue, in an artificial and authoritarian way. It hides the global counterexamples which led to their discovery. Heuristic style on the contrary highlights these factors. It emphasises the problem-situation: it emphasises the 'logic' which gave birth to the new concept.

Let us see first how one can introduce in heuristic style the proof-generated concept of uniform convergence, which we discussed above (appendix 1). In this and the other examples, we certainly presume familiarity with the technical terms of the method of proofs and refutations. But this is no more demanding than the usual requirement of familiarity with the technical terms of the Euclidean programme, like axioms, primitive terms, etc.

(a) Uniform convergence

Thesis   The specific version of the Leibnizian principle of continuity; which states that the limit function of any convergent sequence of continuous functions is continuous. (*Primitive Conjecture*)

(b) there is a fallibilist logic of discovery which is the logic of scientific progress. But Popper, who has laid down the basis of *this* logic of discovery, was not interested in the metaquestion of what was the nature of his inquiry and he did not realise that this is neither psychology nor logic, it is an independent discipline, the logic of discovery, heuristic.

[1] Although it has to be admitted that they would be much fewer too, as the statement of the problem-situation would too obviously display the pointlessness of quite a few of them.

Antithesis Cauchy's definition of continuity raises the thesis to a higher level. His *definitional decision* legalises Fourier's counterexamples. This definition at the same time excludes the possible compromise that continuity be restored by perpendicular lines, and so gives rise – together with some trigonometrical series – to the negative pole of the antithesis. The 'positive pole' gets strengthened by Cauchy's proof, which will be the proof-ancestor of uniform convergence. The 'negative pole' gets strengthened by more and more *global counterexamples* to the primitive conjecture.

Synthesis The *guilty lemma* to which the global counterexamples are also *local* ones is spotted, the proof improved, the conjecture improved. The characteristic constituents of the synthesis emerge; the *theorem* and with it the *proof-generated concept* of uniform convergence.[1]

The Hegelian language, which I use here, would I think, generally be capable of describing the various developments in mathematics. (It has, however, its dangers as well as its attractions.) The Hegelian conception of heuristic which underlies the language is roughly this.

---

[1] For some reason, uniform convergence is, in some text books, singled out for exceptional (quasi-heuristic) treatment. For instance W. Rudin in his [1953], first introduces a section: 'Discussion of Main Problem' (p. 115), where he proposes the primitive conjecture and its refutation and only then introduces the definition of uniform convergence. This presentation has two blemishes: (*a*) Rudin does not present only the primitive conjecture and its refutation, but rather asks whether the primitive conjecture is true or false, and shows falsehood by the well-known examples. But by doing this he does not go beyond infallibilist style; in his 'problem-situation' there is no conjecture but rather a sharp and sophisticated question, followed by an example (not by a counterexample) which gives the unwavering answer. (*b*) Rudin does not show that the concept of uniform convergence emerges from the proof, rather, in his presentation, the definition precedes the proof. This could not be otherwise in the deductivist style, because if he had given first the original proof, and only then the refutation followed by the improved proof and by the proof-generated definition, he would have displayed the movement of 'eternally static' mathematics, the fallibility of 'infallible' mathematics, which is inconsistent with the Euclidean tradition. (Perhaps it should be added that I keep quoting Rudin's book because it is one of the best textbooks within this tradition.) In the preface, for instance, Rudin says: 'It seems important, particularly for a beginner, to see explicitly that the hypotheses of a theorem are really needed to ensure the validity of the conclusions. For this purpose a fairly large number of counterexamples have been included in the text'. Unfortunately these are mock-counterexamples, as in fact they are examples to show how wise mathematicians have to include all the hypotheses in the theorem. But he does not say where these hypotheses come from, that they come from the proof-ideas, and that the theorem does not jump out of the head of the mathematician, like Pallas Athene, fully armed out of Zeus's head. His use of the word 'counterexample' should not misguide us into expecting a fallibilist style. *Editors' note:* All Lakatos's remarks about Rudin's work are based on the *first edition* of this book. Not all the passages Lakatos quotes are to be found in the second edition, which appeared in 1964.

Mathematical activity is human activity. Certain aspects of this activity – as of any human activity – can be studied by psychology, others by history. Heuristic is not primarily interested in these aspects. But mathematical activity produces mathematics. Mathematics, this product of human activity, 'alienates itself' from the human activity which has been producing it. It becomes a living, growing organism, that *acquires a certain autonomy* from the activity which has produced it; it develops its own autonomous laws of growth, its own dialectic. The genuine creative mathematician is just a personification, an incarnation of these laws which can only realise themselves in human action. Their incarnation, however, is rarely perfect. The activity of human mathematicians, as it appears in history, is only a fumbling realisation of the wonderful dialectic of mathematical ideas. But any mathematician, if he has talent, spark, genius, communicates with, feels the sweep of, and obeys this dialectic of ideas.[1]

Now heuristic is concerned with the autonomous dialectic of mathematics and not with its history, though it can study its subject only through the study of history and through the rational reconstruction of history.[2]

(b) Bounded variation

The way the concept of bounded variation is generally introduced in textbooks of analysis is a beautiful example of authoritarian deductivist style. Let us take Rudin's book again. In the middle of his chapter on

[1] This Hegelian idea of the autonomy of alienated human activity may provide the clue to some problems concerning the status and methodology of social sciences, especially economics. My concept of the mathematician as an imperfect personification of Mathematics is closely analogous to Marx's concept of the capitalist as the personification of Capital. Unfortunately Marx did not qualify his conception by stressing the imperfect character of this personification, and that there is nothing inexorable about the realisation of this process. On the contrary, human activity can always suppress or distort the autonomy of alienated processes and can give rise to new ones. The neglect of this interaction was a central weakness of Marxist dialectic.

[2*] *Editors' note*: We feel sure that Lakatos would have modified this passage in some respects, for the grip of his Hegelian background grew weaker and weaker as his work progressed. He did, however, retain a belief in the central importance of recognising the partial autonomy of the products of human intellectual endeavour. In this world of the objective content of propositions (Popper came to call it the 'third world': see his [1972]), problems exist (caused, for example, by logical inconsistencies between propositions) independently of our recognition of them; hence we may *discover* (rather than invent) intellectual problems. But Lakatos came to believe that these problems did not 'demand' a solution or dictate their own solution; rather, human ingenuity (which may or may not be forthcoming) is required for their solution. This view is presaged in the criticism of Marx in the above footnote.

the Riemann–Stieltjes integral he suddenly introduces the definition of functions of bounded variation.

6.20. *Definition.* Let f be defined on $[a, b]$. We put

$$(37) \qquad V(f) = \text{lub} \sum_{i=1}^{n} |f(x_i) - f(x_{i-1})|,$$

where the lub is taken over all partitions of $[a, b]$. If $V(f)$ is finite, we say that f is of bounded variation on $[a, b]$, and we call $V(f)$ the total variation of f on $[a, b]$.[1]

Why should we be interested in just this set of functions? The deductivist's answer is: 'Wait and see'. So let us wait, follow the exposition, and try to see. The definition is followed by examples designed to give the reader some ideas about the domain of the concept (this, and things like this, make Rudin's book outstandingly good within the deductivist tradition). Then a series of theorems (6.22, 6.24, 6.25) follows; and then suddenly the following proposition:

Corollary 2. If f is of bounded variation and g is continuous on $[a, b]$, then $f \in \Re^\star(g)$.[2]

($\Re^\star(g)$ is the class of Riemann–Stieltjes functions integrable with respect to g.)

We might be more interested in this proposition if we really understood just why the Riemann–Stieltjes integrable functions are so important. Rudin does not even mention the intuitively most obvious concept of integrability, namely Cauchy-integrability, criticism of which led to Riemann-integrability. So now we have got a theorem in which two mystical concepts, bounded variation and Riemann-integrability, occur. But two mysteries do not add up to understanding. Or perhaps they do for those who have the 'ability and inclination to pursue an abstract train of thought'?[3]

A heuristic presentation would show that both concepts – Riemann–Stieltjes integrability and bounded variation – are proof-generated concepts, originating in one and the same proof: Dirichlet's proof of the Fourier conjecture. This proof gives the problem-background of both concepts.[4]

Now Fourier's primitive conjecture[5] does not contain any mystical

[1] Rudin [1953], pp. 99–100.　　[2] Ibid., p. 106.　　[3] Rudin [1953], Preface.
[4] This proof and the theorem which sums it up are in fact mentioned in Rudin's book, but they are hidden away in exercise 17, of chapter 8 (p. 164), completely disconnected from the above two concepts which are introduced in an authoritarian way.
[5] Fourier [1808], p. 112.

terms. This 'conjecture-ancestor' of bounded variation says that any arbitrary function is Fourier-expandable[1] – which is a simple and most exciting conjecture. The conjecture was proved by Dirichlet.[2] Dirichlet examined his proof carefully and improved Fourier's conjecture by building into it the lemmas as conditions. These conditions are the celebrated Dirichlet conditions. The resulting theorem was this: All functions are Fourier-expandable (1) the value of which at a point of jump is $\frac{1}{2}[f(x+0)+f(x-0)]$, (2) which have only a finite number of discontinuities, and (3) which have only a finite number of maxima and minima.[3]

All these conditions flow from the proof. Dirichlet's proof-analysis was faulty only as regards the third condition: the proof in fact hinges only on the bounded variation of the function. Dirichlet's proof-analysis was criticised and his mistake corrected by C. Jordan in 1881, who thus became the discoverer of the concept of bounded variation. But he did not invent the concept, he did not 'introduce' it[4] – he rather *discovered* it in Dirichlet's proof in the course of a critical re-examination.[5]

Another weakness in Dirichlet's proof was its use of the Cauchy definition of the integral which is a suitable tool only for continuous functions. According to the Cauchy definition, discontinuous func-

---

[1] 'Fourier-expandable' stands for 'expandable into a trigonometrical series with the Fourier-coefficients'.

[2] See his [1829] and [1837]. There are many interesting aspects of the background to this proof we unfortunately cannot now go into; for example, the problem of the value of Fourier's original 'proof', the comparison of the two subsequent Dirichlet-proofs, and Dirichlet's crushing criticism of Cauchy's earlier ([1826]) proof.

[3] It should be mentioned here that Dirichlet's proof was not preceded or stimulated by counterexamples to Fourier's original conjecture. Nobody offered any counterexamples; in fact, Cauchy 'proved' the original conjecture (cf. footnote 2, p. 131; the domain of validity of his proof was the empty set). The first counterexamples were only suggested by the lemmas of Dirichlet's proof; particularly by the first lemma. Apart from this the first counterexample to Fourier's conjecture was presented only in 1876 by du Bois-Reymond, who found a continuous function which was not Fourier-expandable. (du Bois-Reymond [1876].)

[4] To 'introduce' a concept out of the blue is a magical operation which is resorted to very often in history written in deductivist style!

[5] See Jordan [1881] and Jordan [1893], p. 241. Jordan himself stresses that he does not modify Dirichlet's *proof*, but only his *theorem*. ('...Dirichlet's demonstration is thus applicable without modification to every function where oscillation is limited...'). Zygmund, however, is mistaken when stating that Jordan's theorem is 'only more general in appearance' than Dirichlet's (Zygmund [1935], p. 25). This is true of Jordan's proof but not of his theorem. But at the same time it is misleading to say that Jordan 'extended' Dirichlet's theorem to the more general domain of functions with bounded variation. (E.g. Szökefalvi-Nagy [1954], p. 272.) Also Carslaw shows similar lack of understanding of proof-analysis in his *Historical Introduction* to his [1930]. He does not notice that Dirichlet's proof is the proof-ancestor of the proof-generated concept of bounded variation.

tions are not integrable at all, and *ipso facto* they are not Fourier-expandable. Dirichlet avoided this difficulty by regarding the integral of a discontinuous function as the sum of the integrals on those intervals on which the function was continuous. This can easily be done if the number of discontinuities is finite, but leads to difficulties if it is infinite. This is why Riemann criticised Cauchy's concept of integral and invented a new one.

So the two mysterious definitions of bounded variation and of the Riemann-integral are *entzaubert*, deprived of their authoritarian magic; their origin can be traced to some clear-cut problem situation and to the criticism of previous attempted solutions of these problems. The first definition is a proof-generated definition tentatively formulated by Dirichlet and in the end discovered by C. Jordan, critic of Dirichlet's proof-analysis. The second definition comes from the criticism of a previous definition of the integral which turned out to be inapplicable to more complicated problems.

In this second example of heuristic exposition we followed the Popperian pattern of the logic of conjectures and refutations. This pattern follows history more closely than the Hegelian one, which dismisses 'trial and error' as a sheer fumbling human realisation of the necessary development of objective ideas. But even in a rational heuristic of the Popperian brand one has to differentiate between problems which one sets out to solve and problems which one in fact solves; one has to differentiate between 'accidental' errors on the one hand which just disappear, and the criticism of which does not play any role in the further development, and 'essential' errors, which in a sense will be preserved also after refutation and on the criticism of which further development is based. In the heuristic presentation the accidental errors can be omitted without loss, to deal with them is the business of history only.

We have only sketched the first four stages of the proof-procedure which has led to the concept of bounded variation. Here we merely hint at the rest of the intriguing story. The fifth stage,[1] the hunt for the newly found proof-generated concept in other proofs, immediately led to the discovery of bounded variation in the proof of the primitive conjecture that 'all curves are rectifiable'.[2] The seventh stage leads us to the Lebesgue-integral and to modern measure theory.

---

[1] For the list of the standard stages of the method of proofs and refutations, cf. pp. 127–8.

[2] In this discovery again, du Bois-Reymond was a forerunner ([1879], [1885]), and again, the admirably sharp C. Jordan the actual discoverer (Jordan [1887], p. 594–8 and [1893], p. 100–8).

Historical Note. Some heuristically interesting details may be added to the story told in the text. Dirichlet was convinced that the local counterexamples to his second and third lemmas were *not global*; he was convinced that e.g. all continuous functions, regardless of the number of their maxima and minima, are Fourier-expandable. He also hoped that this more general result could be proved by simple *local* amendments in his proof. This idea, that (1) Dirichlet's proof was only a partial one and (2) the final proof could be achieved by some minor amendments, was widely accepted from 1829 to 1876 when du Bois-Reymond produced the *first* genuine counterexample to Fourier's old conjecture and thereby destroyed the hopes for such an amendment. Jordan's discovery of bounded variation seems to have been stimulated by this counter-example.

It is interesting to note that Gauss, too, encouraged Dirichlet to improve his proof so that it should apply to functions with any number of maxima and minima. It is intriguing that although Dirichlet did not solve this problem, either in 1829 or in 1837, still in 1853 he thought the solution to be so obvious that in his letter replying to Gauss's request, he improvised it (Dirichlet [1853]). The gist of his solution is this. The condition that the set of maxima and minima should not have any point of condensation in the interval considered, is in fact a sufficient condition for his proof. That his *second* condition, about the finite number of discontinuities can be amended, was stated by him already in his first 1829 paper. He asserted there that his proof in fact applies only if the set of discontinuities is nowhere dense. These corrections show that Dirichlet was very much concerned with the problem of the analysis of his proof, and was convinced that it applies to more functions than those which satisfy his cautious conditions, later called 'Dirichlet conditions'. It is characteristic that in his [1837] he does not state the theorem at all. He was always convinced that his theorem held for all continuous functions as his letter to Gauss shows and as he himself told the probably sceptical Weierstrass. (Cf. *Ostwald's Klassiker der Exakten Wissenschaften*, 186, 1913, p. 125.)

Now the theorem as stated by him in his [1829] in fact embraces all types of functions 'which occur in nature'. Further, more refined analysis already leads into the realm of very 'pure' analysis. I claim that the analysis of Dirichlet's proof – first of all by Riemann – was the starting point of modern abstract analysis and I find the recently widely accepted view of P. Jourdain about Fourier's decisive role exaggerated. Fourier was not interested in mathematical arguments which went beyond direct applicability. Dirichlet's thinking was indeed different. He vaguely felt that the analysis of his proof required a new conceptual framework. The last sentence of his [1829] paper is a veritable prophecy:

But the thing to be done with all the clarity that one can desire, requires some details bound up with the fundamental principles of the infinitesimal calculus, which will be presented in another note...

But he never published the promised note. It was Riemann who, by criticising the Cauchy concept of the integral, clarified these 'details bound up with the fundamental principles of the infinitesimal calculus', and who, by articulating Dirichlet's vague feelings, and by introducing a revolutionary technique, carried mathematical analysis and, indeed, rationality into the domain of functions which do not occur in nature and which had hitherto been regarded as monsters, or, at best, uninteresting exceptions or 'singularities'. (This was Dirichlet's attitude, expressed in his [1829] paper and in his letter to Gauss [1853].)

Some infallibilist historians of mathematics use here the ahistorical technique of condensing a long development full of struggle and criticism into one single action of infallible insight and attribute to Dirichlet the maturity of later analysts. These antihistorical historians attribute our modern general concept of a real function to Dirichlet, and accordingly name this concept the Dirichlet concept of function. E. T. Bell asserts in his [1945], p. 293 that 'P. G. L. Dirichlet's definition of a (numerical valued) function of a (real, numerical valued) variable as a table, or correspondence, or correlation, between two sets of numbers hinted at a theory of equivalence of point sets'. Bell gives as reference: 'Dirichlet: *Werke*, I, p. 135'. But there is nothing like this there. Bourbaki says: 'It is known that it was on this occasion that Dirichlet, making precise Fourier's ideas, defined the general notion of a function as we understand it today'. (Bourbaki [1960], p. 247.) 'It is known' says Bourbaki, but does not give any reference. We find the remark that this concept of real function 'is due to Dirichlet' in most classical textbooks (e.g. Pierpont [1905], p. 120). Now there is no such definition in Dirichlet's works at all. But there is ample evidence that he had no idea of this concept. In his [1837] paper for instance, when he discusses piecewise continuous functions, he says that at points of discontinuity the function *has two values*:

The curve, whose x and y coordinates are denoted by β and $\phi(\beta)$ respectively, consists of several pieces. At points above the x axis corresponding to certain particular values of β, successive portions of the curve are disconnected; and for each such x co-ordinate there correspond in fact 2 y co-ordinates, of which one belongs to the portion that ends at that point, and the other belongs to the portion that begins there. In what follows it will be necessary to distinguish these two values of $\phi(\beta)$ and we shall denote them by $\phi(\beta - 0)$ and $\phi(\beta + 0)$.

These quotations show beyond any reasonable doubt how far Dirichlet was from the 'Dirichlet concept of function'.

Those who associate Dirichlet with the 'Dirichlet definition' usually think of the Dirichlet function which occurs on the last page of his [1829] paper: a function which is 0 where x is rational and 1 where x is irrational. The trouble again is that Dirichlet still held that all genuine functions are in fact Fourier-expandable – he devised this 'function' explicitly as a monster. According to Dirichlet his 'function' is an example not of an 'ordinary' real function, but of a function which does not really deserve the name.

It is intriguing that those who managed to notice the Dirichlet definition of function despite its absence, did not notice the titles of his two papers, which refer to the expansion of any 'completely arbitrary' (*ganz willkürliche*) functions into Fourier series. But this means that – according to Dirichlet – the Dirichlet function was outside this family of 'completely arbitrary functions', that he regarded it as a monster, because an 'ordinary' function has to have an integral and this obviously had none. Riemann, in fact, criticised Dirichlet's narrow concept of a function when criticising Cauchy's concept of integral together with its *ad hoc* amendment by Dirichlet. Riemann showed that if we widen the concept of integral, a monster like a function which is discontinuous for every rational number of the form $p/2n$, where p is an odd number, prime to n, is integrable, although it is discontinuous on an everywhere dense set. Consequently this function, so akin to Dirichlet's monster, is ordinary. (There was nothing 'arbitrary' in Riemann's extension of the integral concept; his revolutionary step was to ask what kind of functions are represented by trigonometric series, instead of asking what kind of functions are Fourier-expandable. His aim was to expand the concept of integral so much that all functions which are the sums of trigonometrical series should be integrable and thereby Fourier-expandable. This is a most beautiful example of conceptual instrumentalism.)

Perhaps the originator of the tale about Dirichlet's having set up the 'Dirichlet definition of function' should be identified here. It was H. Hankel, who in analysing the development of the concept of function ([1882], pp. 63–112), explained how Fourier's results broke down the old concept of a function; and then, he went on:

It only remained first, to drop the condition that the function should be analytic, on the grounds that such a condition is without significance, and, secondly, while cutting that knot, to give the following explication. A function is called y of x if to each value of the variable x within a certain interval, there corresponds a definite value of y, and this irrespective of whether y depends on x according to the same law throughout the whole interval, and of whether this dependence can be expressed by means of mathematical operations. This purely nominal definition I shall ascribe to Dirichlet because it lies at the basis of his work on Fourier series, which demonstrated the untenability of that older concept...

(c) *The Carathéodory definition of measurable set*

The change from the deductivist to the heuristic approach will certainly be difficult, but some of the teachers of modern mathematics already realise the need for it. Let us look at an example. In modern textbooks on measure theory or probability theory we frequently get confronted by the Carathéodory definition of measurable set:

Let $\mu^\star$ be an outer measure on an hereditary σ-ring **H**. A set E in **H** is $\underline{\mu^\star}$ **measurable** if, for every set A in **H**,

$$\mu^\star(A) = \mu^\star(A \cap E) + \mu^\star(A \cap E')[1]$$

[1] Halmos [1950], p. 44.

The definition as it stands is bound to be puzzling. Of course there is always the easy way out: mathematicians define their concepts just as they like. But serious teachers do not take this easy refuge. Nor can they say that just this is the *correct*, *true* definition of measurability and that mature mathematical insight should see it as such. Usually in fact, they give a rather vague hint that we should look to the conclusions later to be drawn from the definition: 'Definitions are dogmas; only the conclusions drawn from them can afford new insight'.[1] So we have to take the definitions on trust and see what happens. Although this has an authoritarian touch, at least it is a sign that the problem has been realised. It is an apology, if still an authoritarian one. Let us quote Halmos's apology for Carathéodory's definition: 'It is rather difficult to get an intuitive understanding of the meaning of μ^*-measurability except through familiarity with its implications, which we propose to develop below.'[2] And then he goes on:

The following comment may, however, be helpful. An outer measure is not necessarily a countably, nor even finitely, additive set function. In an attempt to satisfy the reasonable requirement of additivity, we single out those sets which split every other set additively – the definition of μ^*-measurability is the precise formulation of this rather loose description. The greatest justification of this apparently complicated concept is, however, its possibly surprising but absolutely complete success as a tool in proving the important and useful extension theorem of §13.[3]

Now the first, 'intuitive', part of this justification is a bit misleading, because, as we learn from the second part, this concept is a proof-generated concept in Carathéodory's theorem about the extension of measures (which Halmos introduces only in the next chapter). So whether it is intuitive or not is not at all interesting: its rationale lies not in its intuitiveness, but in its proof-ancestor. One should never tear a proof-generated definition off from its proof-ancestor and present it sections or even chapters before the proof to which it is heuristically secondary.

M. Loeve, in his [1955] presents the definition very properly in his section on the extension of measures, as a notion needed in the extension theorem: 'We shall need various notions that we collect here.'[4] But how on earth can he know which of these most complicated instruments will be needed for the operation? Certainly he already has

---

[1] K. Menger [1928], p. 76, quoted with approval by K. R. Popper in his [1959], p. 55.
[2] Halmos [1950], p. 44.   [3] Ibid.
[4] Loève [1955], p. 87.

some idea what he will find and how he will proceed. But why then, this mystical set-up of putting the definition before the proof?

One can easily give more examples, where stating the primitive conjecture, showing the proof, the counterexamples, and following the heuristic order up to the theorem and to the proof-generated definition would dispel the authoritarian mysticism of abstract mathematics, and would act as a brake on degeneration. A couple of case-studies in this degeneration would do much good for mathematics. Unfortunately the deductivist style and the atomisation of mathematical knowledge protect 'degenerate' papers to a very considerable degree.

BIBLIOGRAPHY

(Revised and extended by Gregory Currie)

Abel, N. H. [1825] 'Letter to Holmboë', in S. Lie and L. Sylow (eds.): *Oeuvres Complètes*, vol. 2. Christiana: Grøndahl, 1881, pp. 257–8.

Abel, N. H. [1826*a*] 'Letter to Hansteen', in S. Lie and L. Sylow (eds.): *Oeuvres Complètes*, vol. 2. Christiania: Grøndahl, 1881, pp. 263–5.

Abel, N. H. [1826*b*] 'Untersuchungen über die Reihe

$$1 + \frac{m}{1}x + \frac{m.(m-1)}{2}x^2 + \frac{m.(m-1)(m-2)}{2.3}x^3\ldots',$$

Journal für die Reine und Angewandte Mathematik, **1**, pp. 311–39.

Abel, N. H. [1881] 'Sur les Séries', in S. Lie and L. Sylow (eds.): *Oeuvres Complètes*, vol. 2. Christiania: Grøndahl, pp. 197–205.

Aetius [*c.* 150] *Placita*, in H. Diels (ed.): *Doxographi Graeci*. Berolini: Reimeri, 1879.

Aleksandrov, A. D. [1956] 'A General View of Mathematics', in A. D. Aleksandrov, A. N. Kolmogorov and M. A. Lavrent'ev (eds.): *Mathematics: its Content, Methods and Meaning*. (English translation by S. H. Gould, K. A. Hirsch and T. Bartha. Cambridge, Massachusetts: M.I.T. Press, 1963).

Ambrose, A. [1959] 'Proof and the Theorem Proved', *Mind*, **68**, pp. 435–45.

Arber, A. [1954] *The Mind and the Eye*. Cambridge: Cambridge University Press.

Arnauld, A. and Nicole, P. [1724] *La Logique, ou L'Art de Penser*. Lille: Publications de la Faculté des Lettres et Sciences Humaines de l'Université de Lille, 1964.

Bacon, F. [1620] *Novum Organum*. English translation in R. L. Ellis and J. Spedding (eds.): *The Philosophical Works of Francis Bacon*. London: Routledge, 1905, pp. 241–387.

Baltzer, R. [1862] *Die Elemente der Mathematik*, vol. 2. Leipzig: Hirzel.

Bartley, W. W. [1962] *Retreat to Commitment*. New York: Alfred A. Knopf.

Becker, J. C. [1869*a*] 'Über Polyeder', *Zeitschrift für Mathematik und Physik*, **14**, pp. 65–76.

Becker, J. C. [1869*b*] 'Nachtrag zu dem Aufsätze über Polyeder', *Zeitschrift für Mathematik und Physik*, **14**, pp. 337–343.

Becker, J. C. [1874] 'Neuer Beweis und Erweiterung eines Fundamentalsatzes über Polyederflächen', *Zeitschrift für Mathematik und Physik*, **19**, pp. 459–60.

Bell, E. T. [1945] *The Development of Mathematics*. Second edition. New York: McGraw-Hill.

Bérard, J. B. [1818–19] 'Sur le Nombre des Racines Imaginaires des Équations; en Réponse aux Articles de MM. Tédenat et Servois', *Annales de Mathématiques, Pures et Appliquées*, **9**, pp. 345–72.

Bernays, P. [1947] Review of Pólya [1945], *Dialectica* **1**, pp. 178–88.

Bolzano, B. [1837] *Wissenschaftslehre*. Leipzig: Meiner, 1914–31.

Bourbaki, N. [1949] *Topologie Général*. Paris: Hermann.

Bourbaki, N. [1960] *Éléments d'Histoire des Mathématiques*. Paris: Hermann.

Boyer, C. [1939] *The Concepts of the Calculus*. New York: Dover, 1949.

Braithwaite, R. B. [1953] *Scientific Explanation*. Cambridge: Cambridge University Press.

Brouwer, L. E. J. [1952] 'Historical background, Principles and Methods of Intuitionism', *South African Journal of Science*, **49**, pp. 139–46.

Carnap, R. [1937] *The Logical Syntax of Language*. New York and London: Kegan Paul. (Revised translation of *Logische Syntax der Sprache*, Vienna: Springer, 1934.)

Carslaw, H. S. [1930] *Introduction to the Theory of Fourier's Series and Integrals*. Third edition. New York: Dover, 1950.

Cauchy, A. L. [1813a] 'Recherches sur les Polyèdres', *Journal de l'École Polytechnique*, **9**, pp. 68–86. (Read in February 1811).

Cauchy, A. L. [1813b] 'Sur les Polygones et les Polyèdres', *Journal de l'École Polytechnique*, **9**, pp. 87–98. (Read in January 1812.)

Cauchy, A. L. [1821] *Cours d'Analyse de l'École Royale Polytechnique*. Paris: de Bure.

Cauchy, A. L. [1826] 'Mémoire sur les Développements des Functions en Séries Périodiques', *Mémoires de l'Académie des Sciences* **6**, pp. 603–12.

Cauchy, A. L. [1853] 'Note sur les Séries Convergentes dont les Divers Terms sont des Fonctions Continues d'une Variable Réelle ou Imaginaire entre des Limites Données', *Comptes Rendus Hebdomadaires des Séances de l'Académie des Sciences*, **37**, pp. 454–9.

Cayley, A. [1859] 'On Poinsot's Four New Regular Solids', *The London, Edinburgh, and Dublin Philosophical Magazine and Journal of Science*, 4th Series, **17**, pp. 123–8.

Cayley, A. [1861] 'On the Partitions of a Close', *The London, Edinburgh, and Dublin Philosophical Magazine and Journal of Science*, 4th Series, **21**, pp. 424–8.

Church, A. [1956] *Introduction to Mathematical Logic*, vol. 1. Princeton: Princeton University Press.

Clairaut, A. C. [1741] *Eléments de Géométrie*. Paris: Gauthier-Villars.

Copi, I. M. [1949] 'Modern Logic and the Synthetic *A Priori*', *The Journal of Philosophy*, **46**, pp. 243–5.

Copi, I. M. [1950] 'Gödel and the Synthetic *A Priori*: a Rejoinder', *The Journal of Philosophy*, **47**, pp. 633–6.

Crelle, A. L. [1826–7] *Lehrbuch der Elemente der Geometrie*, vols. 1 and 2, Berlin: Reimer.

Curry, H. B. [1951] *Outlines of a Formalist Philosophy of Mathematics.* Amsterdam: North Holland.

Darboux, G. [1874a] 'Lettre à Houel, 12 Janvier'. (Quoted in F. Rostand: *Souci d'Exactitude et Scrupules des Mathématiciens.* Paris: Librairie Philosophique J. Vrin, 1960, p. 11.)

Darboux, G. [1874b] 'Lettre à Houel, 19 Février'. (Quoted in F. Rostand: *Souci d'Exactitude et Scrupules des Mathématiciens.* Paris: Librairie Philosophique J. Vrin, 1960 p. 194.)

Darboux, G. [1875] 'Mémoire sur les Fonctions Discontinues', *Annales Scientifiques de l'École Normale Supérieure*, second series **4**, pp. 57–112.

Darboux, G. [1883] 'Lettre à Houel, 2 Septembre'. (Quoted in F. Rostand: *Souci d'Exactitude et Scrupules des Mathématiciens.* Paris: Librairie Philosophique J. Vrin, 1960, p. 261.)

Denjoy, A. [1919] 'L'Orientation Actuelle des Mathématiques', *Revue du Mois*, **20**, pp. 18–28.

Descartes, R. [1628] *Rules for the Direction of the Mind.* English translation in E. S. Haldane and G. R. T. Ross (eds.): *Descartes' Philosophical Works*, vol. 1, Cambridge: Cambridge University Press, 1911.

Descartes, R. [1639] *De Solidorum Elementis.* (First published in Foucher de Careil: *Oeuvres Inédites de Descartes*, vol. 2, Paris: August Durand, 1860, pp. 214–34. For a considerably improved text see C. Adam and P. Tannery (eds.): *Oeuvres de Descartes*, vol. 10, pp. 257–78, Paris: Cerf, 1908.)

Dieudonné, J. [1939] 'Les Méthodes Axiomatiques Modernes et les Fondements des Mathématiques', *Revue Scientifique*, **77**, pp. 224–32.

Diogenes Laertius [c. 200] *Vitae Philosophorus.* With an English translation by R. D. Hicks. Vol. 2, London: Heinemann, 1925.

Dirichlet, P. L. [1829] 'Sur la Convergence des Séries Trigonométriques que servent à représenter une Fonction Arbitraire entre les Limites Données', *Journal für die Reine und Angewandte Mathematik*, **4**, pp. 157–69.

Dirichlet, P. L. [1837] 'Über die Darstellung Ganz Willkürlicher Functionen durch Sinus- und Cosinusreihen', in H. W. Dove and L. Moser (eds.): *Repertorium der Physik*, **1**, pp. 152–74.

Dirichlet, P. L. [1853] 'Letter to Gauss, 20 February, 1853', in L. Kronecker (ed.): *Werke*, vol. 2 pp. 385–7. Berlin: Reiner, 1897.

du Bois-Reymond, P. D. G. [1875] 'Beweis, das die Coefficienten der Trigonometrischen Reihe $f(x) = \sum_{p=0}^{p=\infty} (a_p \cos px + b_p \sin px)$ die werte

$$a_0 = \frac{1}{2\pi} \int_{-\pi}^{+\pi} d\alpha\, f(\alpha), \quad a_p = \frac{1}{\pi} \int_{-\pi}^{+\pi} d\alpha\, f(\alpha) \cos p\alpha,$$

$$b_p = \frac{1}{\pi} \int_{-\pi}^{+\pi} d\alpha\, f(\alpha) \sin p\alpha$$

haben, jedesmal wenn diese Integrale Endlich und Bestimmt sind', *Abhandlungen der Königlich-Bayerischen Akademie der Wissenschaften, Mathematisch-Physikalischen Classe*, **12**, 1, pp. 117–166.

du Bois-Reymond, P. D. G. [1876] 'Untersuchungen über die Convergenz und Divergenz der Fourier'schen Darstellungsformeln', *Abhandlungen der Königlich-Bayerischen Akademie der Wissenschaften, Mathematisch-Physikalischen Classe*, **12**, 2, pp. i–xxiv and 1–102.

du Bois-Reymond, P. D. G. [1879] 'Erläuterungen zu den Anfangsgründen der Variationrechnung', *Mathematische Annalen*, **15**, pp. 282–315, 564–76.

du Bois-Reymond, P. D. G. [1885] Über den Begriff der Länge einer Curve', *Acta Mathematica*, **6**, pp. 167–8.

Dyck, W. [1888] 'Beiträge zur Analysis Situs', *Mathematische Annalen*, **32**, pp. 457–512.

Einstein, A. [1953] 'Letter to P. A. Schilpp'. Published in P. A. Schilpp: 'The Abdication of Philosophy', *Kant Studien*, **51**, pp. 490–1, 1959–60.

Euler, L. [1756–7] 'Specimen de usu Observationum in Mathesi Pura', *Novi Comentarii Academiae Scientiarum Petropolitanae*, **6**, pp. 185–230. Editorial summary, pp. 19–21.

Euler, L. [1758a] 'Elementa Doctrinae Solidorum', *Novi Commentarii Academiae Scientiarum Petropolitanae*, **4**, pp.109–40. (Read in November 1750.)

Euler, L. [1758b] 'Demonstratio Nonnullarum Insignium Proprietatus Quibus Solida Hedris Planis Inclusa sunt Praedita', *Novi Commentarii Academiae Scientiarum Petropolitanae*, **4**, pp. 140–60. (Read in September 1751.)

Eves, H. and Newsom, C. V. [1958] *An Introduction to the Foundations and Fundamental Concepts of Mathematics*. New York: Rinehart.

Félix, L. [1957] *L'Aspect Moderne des Mathématiques*. (English translation by J. H. Hlavaty and F. H. Hlavaty: *The Modern Aspect of Mathematics*, New York: Basic Books, 1960.)

Forder, H. G. [1927] *The Foundations of Euclidean Geometry*. New York: Dover, 1958.

Fourier, J. [1808] 'Mémoire sur la Propagation de la Chaleur dans les Corpe Solides (Extrait)', *Nouveau Bulletin des Sciences, par la Société Philomathique de Paris*, **1**, pp. 112–16.

Fréchet, M. [1928] *Les Éspaces Abstraits*. Paris: Gauthier-Villars.

Fréchet, M. [1938] 'L'Analyse Générale et la Question des Fondements', in F. Gonseth (ed.): *Les Entretiens de Zürich, sur les Fondements et la Méthode des Sciences Mathématiques*, Zürich: Leemans Frères et Cie, 1941. pp. 53–73.

Frege, G. [1893] *Grundgesetze der Arithmetik*, vol. 1, Hildesheim: George Olms, 1962.

Gamow, G. [1953] *One, Two, Three...Infinity*. New York: The Viking Press.

Gauss, C. F. [1813] 'Disquisitiones Generales Circa Seriem Infinitam

$$1 + \frac{\alpha\beta}{1.\gamma}.x + \frac{\alpha(\alpha+1)\beta(\beta+1)}{1.2.\gamma(\gamma+1)}x.x + \frac{\alpha(\alpha+1)(\alpha+2)\beta(\beta+1)(\beta+2)}{1.2.3.\gamma(\gamma+1)(\gamma+2)}.x^3 + \text{etc.',}$$

in *Werke*, vol. 3, pp. 123–62. Leipzig: Teubner.

Gergonne, J. D. [1818] 'Essai sur la Théorie des Definitions', *Annales de Mathématiques, Pures et Appliquées*, 9, pp. 1–35.

Goldschmidt, R. [1933] 'Some Aspects of Evolution', *Science*, 78, pp. 539–47.

Grunert, J. A. [1827] 'Einfacher Beweis der von Cauchy und Euler Gefundenen Sätze von Figurennetzen und Polyedern', *Journal für die Reine und Angewandte Mathematik*, 2, p. 367.

Halmos, P. [1950] *Measure Theory*. New York and London: Van Nostrand Reinhold.

Hankel, H. [1882] 'Untersuchungen über die Unendlich oft Oscillierenden und Unstetigen Functionen', *Mathematische Annalen*, 20, pp. 63–112.

Hardy, G. H. [1918] 'Sir George Stokes and the Concept of Uniform Convergence', *Proceedings of the Cambridge Philosophical Society*, 19, pp. 148–56.

Hardy, G. H. [1928] 'Mathematical Proof', *Mind*, 38, pp. 1–25.

Haussner, R. (ed.) [1906] *Abhandlungen über die Regelmassigen Sternkörper*. Ostwald's Klassiker der Exacten Wissenschaften, No. 151, Leipzig: Engelmann.

Heath, T. L. [1925] *The Thirteen Books of Euclid's Elements*. Second edition. Cambridge: Cambridge University Press.

Hempel, C. G. [1945] 'Studies in the Logic of Confirmation, 1 and 2', *Mind*, 54. pp. 1–26 and 97–121.

Hermite, C. [1893] 'Lettre à Stieltjes, 20 Mai 1893', in B. Baillaud and H. Bourget (eds.): *Correspondence d'Hermite et de Stieltjes*, 2, pp. 317–19. Paris: Gautheirs-Villars, 1905.

Hessel, J. F. [1832] 'Nachtrag zu dem Euler'schen Lehrsatze von Polyedern', *Journal für die Reine und Angewandte Mathematik*, 8, pp. 13–20.

Heyting, A. [1939] 'Les Fondements des Mathématiques du Point de Vue Intuitionniste', in F. Gonseth: *Philosophie Mathématique*, Paris: Hermann, pp. 73–5.

Heyting, A. [1956] *Intuitionism: An Introduction*. Amsterdam: North Holland.

Hilbert, D. and Cohn-Vossen, S. [1932] *Anschauliche Geometrie*. Berlin: Springer. English translation by P. Nemenyi: *Geometry and the Imagination*. New York: Chelsea (1956).

Hobbes, T. [1651] *Leviathan*, in W. Molesworth (ed.): *The English Works of Thomas Hobbes*, vol. 3. London: John Bohn, 1839.

Hobbes, T. [1656] *The Questions Concerning Liberty, Necessity and Chance*, in W. Molesworth (ed.): *The English Works of Thomas Hobbes*, vol. 5. London: John Bohn, 1841.

Hölder, O. [1924] *Die Mathematische Methode*. Berlin: Springer.

Hoppe, R. [1879] 'Ergänzung des Eulerschen Satzes von den Polyedern', *Archiv der Mathematik und Physik*, **63**, pp. 100–3.

Husserl, E. [1900] *Logische Untersuchungen*, vol. 1. Tubingen: Niemeyer, 1968.

Jonquières, E. de [1890a] 'Note sur un Point Fondamental de la Théorie des Polyèdres', *Comptes Rendus des Séances de l'Académie des Sciences*, **110**, pp. 110–15.

Jonquières, E. de [1890b] 'Note sur le Théorème d'Euler dans la Théorie des Polyèdres', *Comptes Rendus des Séances de l'Académie des Sciences*, **110**, pp. 169–73.

Jordan, C. [1866a] 'Recherches sur les Polyèdres', *Journal für die Reine und Angewandte Mathematik*, **66**, pp. 22–85.

Jordan, C. [1866b] 'Résumé de Recherches sur la Symétrie des Polyèdres non Eulériens', *Journal für die Reine und Angewandte Mathematik*, **66**, pp. 86–91.

Jordan, C. [1881] 'Sur la Série de Fourier', *Comptes Rendus des Séances de l'Académie des Sciences*, **92**, pp. 228–33.

Jordan, C. [1887] *Cours d'Analyse de l'École Polytechnique*, vol 3, first edition. Paris: Gauthier-Villars.

Jordan, C. [1893] *Cours d'Analyse de l'École Polytechnique*, vol. 1, second edition. Paris: Gauthier-Villars.

Jourdain, P. E. B. [1912] 'Note on Fourier's Influence on the Conceptions of Mathematics', *Proceedings of the Fifth International Congress of Mathematics*, **2**, pp. 526–7.

Kant, I. [1781] *Critik der Reinen Vernunft*. First edition.

Kepler, I. [1619] *Harmonice Mundi*, in M. Caspar and W. von Dyck (eds.): *Gesammelte Werke*, vol. 6. Munich: C. H. Beck, 1940.

Knopp, K. [1928] *Theory and Application of Infinite Series*. (Translated by R. C. Young, London and Glasgow: Blackie, 1928.)

Lakatos, I. [1961] *Essays in the Logic of Mathematical Discovery*, unpublished Ph.D. Dissertation, Cambridge.

Lakatos, I. [1962] 'Infinite Regress and the Foundations of Mathematics', *Aristotelian Society Supplementary Volumes*, **36**, pp. 155–84.

Lakatos, I. [1970] 'Falsification and the Methodology of Scientific Research Programmes', in I. Lakatos and A. E. Musgrave (eds.): *Criticism and the Growth of Knowledge*, Cambridge: Cambridge University Press, pp. 91–196.

Landau, E. [1930] *Grundlagen der Analysis*. Leipzig: Akademische Verlagsgesellschaft.

Lebesgue, H. [1923] 'Notice sur la Vie et les Travaux de Camille Jordan', *Mémoires de l'Académie de l'Institute de France*, **58**, pp. 34–66. Reprinted in H. Lebesgue, *Notices d'Histoire des Mathématiques*, Genève. pp. 40–65.

Lebesgue, H. [1928] *Leçons sur l'Intégration et la Recherche des Fonctions Primitives*. Paris: Gauthier-Villars. (Second, enlarged edition of the original 1905 version.)

Legendre, A.-M. [1809] *Éléments de Géométrie*. Eighth edition. Paris: Didot. The first edition appeared in 1794.

Leibniz, G. W. F. [1687] 'Letter to Bayle', in C. I. Gerhardt (ed.): *Philosophische Schriften*, vol. 3. Hildesheim: George Olms (1965), p. 52.

Lhuilier, S. A. J. [1786] *Exposition Élémentaire des Principes des Calculs Supérieurs*. Berlin: G. J. Decker.

Lhuilier, S. A. J. [1812–13a] 'Mémoire sur la Polyèdrométrie', *Annales de Mathématiques, Pures et Appliquées*, 3, pp. 168–91.

Lhulier, S. A. J. [1812–13b] 'Mémoire sur les Solides Réguliers', *Annales de Mathématiques, Pures et Appliquées*, 3, pp. 233–7.

Listing, J. B. [1861] 'Der Census Räumlicher Complexe', *Abhandlungen der Königlichen Gesellschaft der Wissenschaften zu Göttingen*, 10, pp. 97–182.

Loève, M. [1955] *Probability Theory*. New York: Van Nostrand.

Matthiessen, L. [1863] 'Über die Scheinbaren Einschränkungen des Euler'schen Satzes von den Polyedern', *Zeitschrift für Mathematik und Physik*, 8, pp. 449–50.

Meister, A. L. F. [1771] 'Generalia de Genesi Figurarum Planarum et inde Pendentibus Earum Affectionibus', *Novi Commentarii Societatis Reglae Scientiarum Gottingensis*, 1, pp. 144–80.

Menger, K. [1928] *Dimensionstheorie*. Berlin: Teubner.

Möbius, A. F. [1827] *Der Barycentrische Calcul*. Hildesheim: George Olms, 1968.

Möbius, A. F. [1865] 'Über die Bestimmung des Inhaltes eines Polyeders', *Berichte Königlich-Sächsischen Gesellschaft der Wissenschaften, Mathematisch-Physikalische Classe*, 17, pp. 31–68.

Moigno, F. N. M. [1840–1] *Leçons de Calcul Differentiel et de Calcul Intégral*, 2 vols. Paris: Bachelier.

Moore, E. H. [1902] 'On the Foundations of Mathematics', *Science*, 17, pp. 401–16.

Morgan, A. de [1842] *The Differential and Integral Calculus*. London: Baldwin and Gadock.

Munroe, M. E. [1953] *Introduction to Measure and Integration*. Cambridge, Massachusetts: Addison-Wesley.

Neumann, J. von [1947] 'The Mathematician', in Heywood, R. B. (ed.): *The Works of the Mind*. Chicago: Chicago University Press.

Newton, I. [1717] *Opticks*. Second Edition. London: Dover, 1952.

Olivier, L. [1826] 'Bemerkungen über Figuren, die aus Behebigen, von Geraden Linien Umschlossenen Figuren Zusammengesetzt sind', *Journal für die Reine und Angewandt Mathematik*, 1, pp. 227–31.

Pascal, B. [1659] *Les Réflexions sur la Géométrie en Général (De l'Ésprit Géométrique et de l'Art de Persuader)*. In J. Chevalier (ed.): *Oeuvres Complètes*, Paris: La Librairie Gallimard, 1954, pp. 575–604.

Peano, G. [1894] *Notations de Logique Mathématique*. Turin: Guadagnini.

Pierpont, J. [1905] *The Theory of Functions of Real Variables*, vol. 1. New York: Dover, 1959.

Poincaré, H. [1893] 'Sur la Généralisation d'un Théorème d'Euler relatif aux Polyèdres', *Comptes Rendus de Séances de l'Académie des Sciences*, 117, p. 144.

Poincaré, H. [1899] 'Complément à l'Analysis Situs', *Rendiconti del Circolo Matematico di Palermo*, 13, pp. 285–343.

Poincaré, H. [1902] *La Science et l'Hypothèse*. Paris: Flammarion. Authorised English translation by G. B. Halsted: *The Foundations of Science*, Lancaster, Pennsylvania: The Science Press, 1913, pp. 27–197.

Poincaré, H. [1905] *La Valeur de la Science*. Paris: Flammarion. Authorised English translation by G. B. Halsted: *The Foundations of Science*, Lancaster, Pennsylvania: The Science Press, 1913, pp. 359–546.

Poincaré, H. [1908] *Science et Méthode*. Paris: Flammarion. Authorised English translation by G. B. Halsted: *The Foundations of Science*, Lancaster, Pennsylvania: The Science Press, pp. 546–854.

Poinsot, L. [1810] 'Mémoire sur les Polygones et les Polyèdres', *Journal de l'École Polytéchnique*, 4, pp. 16–48. Read in July 1809.

Poinsot, L. [1858] 'Note sur la Théorie des Polyèdres', *Comptes Rendus de l'Académie des Sciences*, 46, pp. 65–79.

Pólya, G. [1945] *How to Solve It*. Princeton: Princeton University Press.

Pólya, G. [1954] *Mathematics and Plausible Reasoning*, vols. 1 and 2. London: Oxford University Press.

Pólya, G. [1962a] *Mathematical Discovery*, 1. New York: Wiley.

Pólya, G. [1962b] 'The Teaching of Mathematics and the Biogenetic Law', in I. J. Good (ed.): *The Scientist Speculates*. London: Heinemann, pp. 352–6.

Pólya, G. and Szegö, G. [1927] *Aufgaben und Lehrsätze aus der Analysis*, vol 1. Berlin: Springer.

Popper, K. R. [1934] *Logik der Forschung*. Vienna: Springer.

Popper, K. R. [1935] 'Letter to the Editor', *Erkenntnis*, 3, pp. 426–9. Republished in Appendix *i to Popper [1959], pp. 311–14.

Popper, K. R. [1945] *The Open Society and its Enemies*. 2 volumes, London: Routledge and Kegan Paul.

Popper, K. R. [1947] 'Logic Without Assumptions', *Aristotelian Society Proceedings*, 47, pp. 251–92.

Popper, K. R. [1952] 'The Nature of Philosophical Problems and their Roots in Science', *The British Journal for the Philosophy of Science*, 3, pp. 124–56. Reprinted in Popper [1963a].

Popper, K. R. [1957] *The Poverty of Historicism*. London: Routledge and Kegan Paul.

Popper, K. R. [1959] *The Logic of Scientific Discovery*. English translation of [1934]. London: Hutchinson.

Popper, K. R. [1963a] *Conjectures and Refutations*. London: Routledge and Kegan Paul.

Popper, K. R. [1963b] 'Science: Problems, Aims, Responsibilities', *Federation of American Societies for Experimental Biology: Federation Proceedings,* 22, pp. 961–72.

Popper, K. R. [1972] *Objective Knowledge.* Oxford University Press.

Pringsheim, A. [1916] 'Grundlagen der Allgemeinen Functionenlehre', in M. Burkhardt, W. Wutinger and R. Fricke (eds.): *Encyklopädie der Mathematischen Wissenschaften,* vol. 2. Erste Teil, Erste Halbband, pp. 1–53. Leipzig: Teubner.

Quine, W. V. O. [1951] *Mathematical Logic.* Revised edition. Cambridge, Massachusetts: Harvard University Press.

Ramsey, F. P. [1931] *The Foundations of Mathematics and Other Essays.* Edited by R. B. Braithwaite. London: Kegan Paul.

Raschig, L. [1891] 'Zum Eulerschen Theorem der Polyedrometrie', *Festschrift des Gymnasium Schneeberg.*

Reichardt, H. [1941] 'Losung der Aufgabe 274', *Jarhresberichte der Deutschen Mathematiker-Vereinigung,* 51, p. 23.

Reichenbach, H. [1947] *Elements of Symbolic Logic.* New York: Macmillan.

Reiff, R. [1889] *Geschichte der Unendlichen Reihen.* Tubingen: H. Laupp'schen.

Reinhardt, C. [1885] 'Zu Möbius Polyedertheorie. Vorgelegt von F. Klein', *Berichte über die Verhandlungen der Königlich-Sachsischen Gesellschaft der Wissenschaften zu Leipzig,* 37, pp. 106–25.

Riemann, B. [1851] *Grundlagen der eine Allgemeine Theorie der Functionen einer Veranderlichen Complexen Grösse.* (Inaugural dissertation) In M. Weber and R. Dedekind (eds.): *Gesammelte Mathematische Werke und Wissenschaftlicher Nachlass.* Second edition. Leipzig: Teubner, 1892, pp. 3–48.

Riemann, B. [1868] 'Über die Darstellbarkeit einer Function durch eine Trigonometrische Reihe', *Abhandlungen der Königlichen Gesellschaft der Wissenschaften zu Göttingen,* 13, pp. 87–132.

Robinson, R. [1936] 'Analysis in Greek Geometry', *Mind,* 45, pp. 464–73.

Robinson, R. [1953] *Plato's Earlier Dialectic.* Oxford: Oxford University Press.

Rudin, W. [1953] *Principles of Mathematical Analysis.* First edition. New York: McGraw-Hill.

Russell, B. [1901] 'Recent Work in the Philosophy of Mathematics', *The International Monthly,* 3. Reprinted as 'Mathematics and the Metaphysicians', in his [1918], pp. 59–74.

Russell, B. [1903] *Principles of Mathematics.* London: Allen and Unwin.

Russell, B. [1918] *Mysticism and Logic.* London: Allen and Unwin.

Russell, B. [1959] *My Philosophical Development.* London: Allen and Unwin.

Russell, B. and Whitehead, A. N. [1910–13] *Principia Mathematica.* Vol. 1, 1910; Vol. 2, 1912; Vol. 3, 1913. Cambridge University Press.

Saks, S. [1933] *Théorie de l'Intégrale.* English translation by L. C. Young: *Theory of the Integral.* Second edition. New York: Hafner, 1937.

Schläfli, L. [1852] 'Theorie der Vielfachen Kontinuität'. Published posthumously in *Neue Denkschriften der Allgemeinen Schweizerischen Gesellschaft für die Gesamten Naturwissenschaften*, **38**, pp. 1–237. Zürich, 1901.

Schröder, E. [1862] 'Über die Vielecke von Gebrochener Seitenzahl oder die Bedeutung der Stern-Polygone in der Geometrie', *Zeitschrift für Mathematik und Physik*, **7**, pp. 55–64.

Seidel, P. L. [1847] 'Note über eine Eigenschaft der Reihen, welche Discontinuirliche Functionen Darstellen', *Abhandlungen der Mathematisch-Physikalischen Klasse der Königlich Bayerischen Akademie der Wissenschaften*, **5**, pp. 381–93.

Sextus Empiricus [*c.* 190] *Against the Logicians*. Greek text with an English translation by R. G. Bury. London: Heinemann, 1933.

Sommerville, D. M. Y. [1929] *An Introduction to the Geometry of N Dimensions*. London: Dover, 1958.

Steiner, J. [1826] 'Leichter Beweis eines Stereometrischen Satzes von Euler', *Journal für die Reine und Angewandte Mathematik*, **1**, pp. 364–7.

Steinhaus, H. [1960] *Mathematical Snapshots*. Revised and enlarged edition. New York: Oxford University Press.

Steinitz, E. [1914–31] 'Polyeder und Raumeinteilungen', in W. F. Meyer and H. Mohrmann (eds.): *Encyklopädie der Mathematischen Wissenschaften*, vol. 3, AB. 12. Leipzig: Teubner.

Stokes, G. [1848] 'On the Critical Values of the Sums of Periodic Series', *Transactions of the Cambridge Philosophical Society*, **8**, pp. 533–83.

Szabó, Á. [1958] ' "Deiknymi" als Mathematischer Terminus fur "Beweisen" ', *Maia*, N.S. **10**, pp. 1–26.

Szabó, Á. [1960] 'Anfänge des Euklidischen Axiomensystems', *Archive for the History of Exact Sciences*, **1**, pp. 37–106.

Szökefalvi-Nagy, B. [1954] *Valós Függvények és Függvénysorok*. Budapest: Tankönyvkiadó.

Tarski, A. [1930*a*] 'Über einige Fundamentale Begriffe der Metamathematik', *Comptes Rendus des Séances de la Société des Sciences et des Lettres de Varsovie*, **23**, Cl. III, pp. 22–9. Published in English in J. H. Woodger (ed.) [1956], pp. 30–7.

Tarski, A. [1930*b*] 'Fundamentale Begriffe der Methodologie der Deduktiven Wissenschaften, 1', *Monatshefte für Mathematik und Physik*, **37**, pp. 361–404. Published in English in J. H. Woodger (ed.) [1956], pp. 60–109.

Tarski, A. [1935] 'On the Concept of Logical Consequence'. Published in J. H. Woodger (ed.) [1956], pp. 409–20. This paper was read in Paris in 1935.

Tarski, A. [1941] *Introduction to Logic and to the Methodology of Deductive Sciences*. Second edition. New York: Oxford University Press, 1946. (This is a partially modified and extended version of *On Mathematical Logic and Deductive Method*, published in Polish in 1936 and in German translation in 1937.)

Turquette, A. [1950] 'Gödel and the Synthetic A Priori', *The Journal of Philosophy*, **47**, pp. 125–9.

Waerden, B. L. van der [1941] 'Topologie und Uniformisierung der Riemannschen Flachen', *Berichte über die Verhandlungen der Königlich-Sachsischen Gesellschaft der Wissenschaften zu Leipzig*, **93**, pp. 147–60.

Whewell, W. [1858] *History of Scientific Ideas*. Vol. 1. (Part one of the third edition of *The Philosophy of the Inductive Sciences*.)

Wilder, R. L. [1944] 'The Nature of Mathematical Proof', *The American Mathematical Monthly*, **52**, pp. 309–23.

Woodger, J. M. (ed.) [1956] *Logic, Semantics, Metamathematics*. Oxford: Clarendon Press.

Young, W. H. [1903–4] 'On Non-Uniform Convergence and Term-by-Term Integration of Series', *Proceedings of the London Mathematical Society*, **1**, second series, pp. 89–102.

Zacharias, M. [1914–31] 'Elementargeometrie', in W. F. Meyer and H. Mohrmann (eds.): *Encyklopädie der Mathematischen Wissenschaften*, **3**, Erste Teil, Zweiter Halbband, pp. 862–1176. Leipzig: Teubner.

Zygmund, A. [1935] *Trigonometrical Series*. New York: Chelsea, 1952.

INDEXES

(Compiled by Gregory Currie)

$n(s)$ = footnote(s) e = term explained q = quotation

Important entries have *italicised* numerals. References marked with an *asterisk* refer to the editors' notes.

INDEX OF NAMES

polyhedra (*cont.*)

76; *n*-spheroid, 77; octahedron, 110; one-sided, 83, 98*n*1, 110; open, 97*n*; picture frame, 19, 21, 33, 48, 67, 77–9, 83; primitive, 109; prisms and pyramids, 84*n*2; regular, 6, 84; simple, 34, 84*n*2, 89, 100, 113; simply-connected, 85, 97*n*, 99, 109, 120; spherical, 33; star, 16–17, 31, 33, 42–3, 62, 64, 84*n*2, 91, 97*n*; tabulated values for *F*, *V*, and *E*, 69; torus, 33; triangular, 33; twin-tetrahedron, 15, 26–7, 80, 97, 100–1; *uneigentliche*, 16*n*; with cavities, 79, 91–2*n*, 97

polytopes, 109; *see* polyhedra

power series, 133–5

pragmatism, 54

probability theory, 125*n*, 152

problems, 6; one does not solve the problem one has set out to solve, 90; problem situation, 144*u*; problem to prove, 7*n*, 41, 64*n*; scientific inquiry begins and ends with problems, 105

proof-analysis, ix, 42–56, 132; approximate, 51; as a fermenting agent for refutations, 48; can make proof infallible, 138–9*n**; concludes with a proof, 107*n*3; discovery of, 136–41; domain of, 64; and lemma incorporation, 36; may decrease content, 57; no limitations on the tools of, 107*n*1; perfect, 47; and proof, 50, 81*n*4; reduced to a triviality, 126; and revolution in rigour, 55; rigorous, 47*e*; safe, 58; without proof, 50

proof and refutations, rechristened 'the method of proofs and refutations', 64; the dialectical unity of, 37; the heuristic rules of, 50, 58–9; the method of, 47–57, 50*e*; *see* proofs and refutations

proof procedure, 106*n*3, 143*n*1; *see* proofs and refutations

proofs, 7–9*e*; changing standards of, 104–5; criticism of, 10–12, 132; crystal clear, 52; deeper, 57, 64, 66, 120; different proofs yield different theorems, 65–6; domain of, 64*e*; Euclidean, 107*n*3; final, 63–5, 97; formal, 1, 124–5; inductive, 138; infinite regress of, 40; and meaning, 124; more rigorous and more embracing, 120; perfect versus imperfect, 139; and proof analysis, 81*n*4; proofs ancestor, 153; proving after improving, 106; rules of, 56–7*n**; as stage in method of proofs and refutations, 127; as tests, 29*n*,

74–5; that do not prove, 29, 37, 41; to improve, 11, 29, 37, 42; trivial extensions of, 97; valid, 100*n**; without a conjecture, 78; *see* proof analysis, proof procedure, thought experiment

proofs and refutations: and concepts, 89–90; discovery of, 136–40; logic of, 5; method of, 64, 83–4, 127–8, 140; *see* proof and refutations

proportion, theory of, 125*n*

psychologism: and crystal clear proof, 52; versus objectivity, 51

psychology, and the context of discovery, 143–4*n*; translated into physiology, 125*n*

rationalism, critical, 4, 54, 68

rational reconstructions, 5, 84*n*2

rectifiable curves, 149

refutability, 100; *see* counterexamples, refutationists, refutations

refutationists, 19–20*n*; 84–5

refutations: fermenting agents for proof analysis, 48; heuristic, 94–5; important and unimportant, 86, 98; lack of refutations causes neglect of proof analysis, 49; logical and heuristic, 92–3; proof-generated, 48; their diminishing returns, 98*n*1; theoretical versus naive, 96–9; *see* counterexamples

retransmission of falsity, principle of, 47, 57*n*3, 63

rigour, 42–56; Abel's and Cauchy's concept of, 138*n*2; absolute, 28*n*2, 52; Cauchy–Weierstrass revolution in, 55, 121*n*, 125*n*; connections with more embracing proofs, 120, degrees of, 51, 54; and Fourier's counterexamples, 131; and proof-analysis, 51–2, 55–7; and scepticism, 121; sufficient, 54; *see* proofs

scepticism, 4–5, 102; and linguistic communication, 51; religious, 54; and rigour, 121; sceptic turned into dogmatist, 46–7

set theory, 19–20*n*, 56

simplicity, 65*n*3

social sciences, methodology of, 146*n*

surrender, method of, 13–14*e*

synthesis, 9*n*; as proof thought experiment, 75; *see* analysis

tautologies, theorems as, 124–5

Taylor's theorem, 134*n*1

Printed in the United Kingdom
by Lightning Source UK Ltd.
116328UKS00001B/46-51